LAB MANUAL & WORKBOOK
FOR CSEC

BIOLOGY
S.B.A.s

ഏ Amy Heeraman ଔ

BSc. (Hons.), MSc. (Distinction), Dip.Ed.

Lab Manual and Workbook for CSEC Biology S.B.A.s

Design and Layout by Amy Heeraman

Cover by Amy Heeraman

Illustrations by Amy Heeraman

ISBN 978-976-8271-85-3

First Printing: 2017

Published by Amy Heeraman
Princes Town, Trinidad, W.I.

Table of Contents

PREFACE

The purpose of this book is to provide Biology students with a simple, usable guide and work book for the School Based Assessment (SBA) component of the CXC/ CSEC Biology syllabus. Subject teachers will find this book a useful resource for teaching the 5 SBA lab skills.

It is envisioned that, through the use of this workbook, students can prepare and familiarize themselves with methods, mark schemes and discussion points in advance of the practical sessions. During the labs, ideally double periods/ sessions, the worksheets will allow for quick and timely completion of each activity. Each lab may be submitted individually at the end of a session, thus eliminating the need for a teacher to collect and mark multiple bulky lab books. Once marked, the lab scripts can be returned to the students, for them to review mistakes and shortcomings, make necessary changes and complete them for submission.

This book is divided into two sections. The first outlines in detail the 5 skills required by the syllabus, and provides suggested mark schemes. Meanwhile the second section, the lab guides section, covers the SBA topics identified in the syllabus. Each lab guide includes helpful instructions and reminders/ hints to help the students maximize their marks and improve the quality of their reports. At the end of each experiment there are both discussion questions and notes to guide and explain the experiment.

ACKNOWLEDGEMENTS

I gratefully acknowledge Ms. Indira Chickree for her gracious assistance in reviewing and developing this lab manual and workbook for CSEC Biology SBAs.

LAB SKILLS IN CSEC BIOLOGY

A. GUIDELINES FROM THE CSEC BIOLOGY SYLLABUS

CXC examines Experimental Skills (XS) in two ways:

1. Practical exercises:
   a. School Based Assessment (SBA) or Paper 03/1 – laboratory and field work facilitated by the school teacher, and, at the end of Form 5, by a moderator.
   b. Alternative to the School Based Assessment or Paper 03/2 – for private candidates. A practical examination that incorporates written exercises and practical activities for 3 questions.
2. Question 1 of the Paper 2 – a compulsory data analysis question (similar to a lab experiment).

Writing over exams/ re-sit exams:

You must pass the SBAs! If you <u>pass the SBAs but not the final exam</u>, then you may re-sit exams and:
   a. Not repeat the SBAs - using your scores from before (you only have 2 years to get this done).
   b. Repeat the SBA component, i.e. do over all labs and have your teacher mark them, or,
   c. Register as a private candidate and write the Paper 03/2.

Why do SBAs/ Labs?

Labs are **a form of assessment**, and contribute SBA marks (20%) towards the final grade in the subject. Labs **facilitate learning** through inquiry (observing, asking questions and analysing or interpreting data). Asking questions is the fundamental basis of scientific study. Meaningful questioning is a skill you need to learn. SBAs help you to understand the world around you, your body and other living organisms.

How many labs are needed in Biology?

The CSEC Biology syllabus for examination from May/June 2015 requires **eighteen (18)** labs minimum over a two-year period, covering 5 skills (see table 1 below) and MUST include the following topics:

1. Ecological study
2. Movement at the molecular level (diffusion, osmosis)
3. Photosynthesis/ respiration
4. Food tests
5. Germination
6. Nutrition and diseases
7. Genetics

Included in the 18 labs, is an **investigative project** which is essentially two related/ consecutive labs. It is planned in Form 4 (The Proposal – Plan and Design) and carried out in Form 5 (The Implementation).

Table 1 - How the 5 skills are assessed for SBA marks/ moderation over the 2 years

Skills	No. of times assessed for SBA marks/ moderation	
	Form 4 (Year 1)	Form 5 (Year 2)
1. Manipulation and Measurement	1	1
2. Observation/ Recording/Reporting	1	1
3. Planning and design	1	1
4. Drawing	1	0
5. Analysis and Interpretation	1	1

Each practical (all 18 or more) must develop a skill, and should be marked with the appropriate mark scheme criteria.

LAB SKILLS IN CSEC BIOLOGY

What is moderation? Who is it for?

Moderation is for the benefit of both the subject teacher(s) as well as the students (mainly Form 5s). During moderation a teacher/ person authorized by CXC, who is independent of your school conducts school visits to:

(A) Check all lab books/ student work – both Form 4 and Form 5 – to ensure the stipulated topics and number of labs are completed on time.
(B) Check that the marking of labs is being conducted according to set criteria. (Note: each of the five skills must have been assessed twice in Form 4 and then twice in Form 5)
(C) Observe and mark Form 5 students (or a sample) who will carry out an experiment and write a report. The subject teacher is also present and must mark the students being moderated.

WHAT IS SCIENCE?

Science uses evidence to learn about the natural world; it is a body of knowledge. Scientific knowledge is testable and experiments can be repeated to gain the same results. Science begins with **observations** – often taking data (quantitative or qualitative) related to what you see, hear or smell.

Facts, problems, or situations which you may not understand require more information gathering (research). You can even ask another scientist. Sometimes you may want **to find out yourself** by **planning and designing** an experiment. The steps involved form the **Scientific Method**.

The **Scientific Method** is an organised and systematized effort to gain knowledge that uses observation and experimentation to describe and explain nature or natural phenomenon. **Your plan and design lab may follow the steps in the Scientific method.**

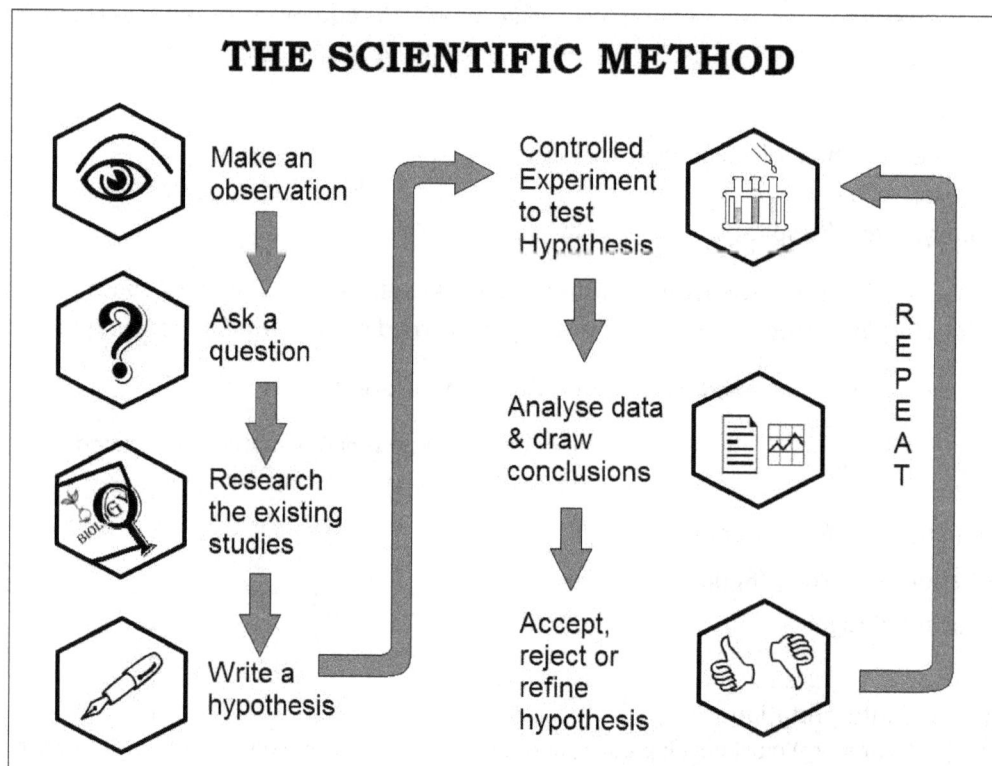

THE SCIENTIFIC METHOD

Make an observation

Ask a question

Research the existing studies

Write a hypothesis

Controlled Experiment to test Hypothesis

Analyse data & draw conclusions

Accept, reject or refine hypothesis

REPEAT

LAB SKILLS IN CSEC BIOLOGY

How Penicillin Was Discovered – The Scientific Method in Practice.

In 1928, Sir Alexander Fleming was studying Staphylococcus bacteria growing in culture dishes (Petri dishes with a growing medium). *Penicillium* a mould (type of fungus) was also observed growing in some of the dishes. Around the circle of mould was a clear area and then an area with bacterial growth. Essentially all the bacteria that was growing close to the mould had died. In the culture dishes without *Penicillium*, no clear areas were present and the bacteria covered the surface.

Fleming decided that the mould was making some chemical that was killing the bacteria. Based on this hypothesis, he decided to extract this chemical from the mould and test it to see if it would kill bacteria. To do this he grew mould in a nutrient broth, a solution that contained all the materials the mould needed to grow. After allowing the mould to grow, he removed it from the nutrient broth. Fleming then took a sample of the nutrient broth and added it to a culture of bacteria, which later died. Using this knowledge he was eventually able to develop antibiotics to treat different diseases.

➢ The problem: A mould *Penicillium* was growing where only bacteria was supposed to be found; this mould was preventing bacteria from growing.
➢ Fleming's hypothesis: Mould was producing a chemical that killed bacteria.
➢ How he tested it: Fleming grew the mould in a liquid, then removed the mould, keeping the liquid. The liquid was then placed on bacteria and then the bacteria died.
➢ The hypothesis was supported based on the observations, and was accepted.
➢ This experiment led to the development of a major medical advancement – antibiotics!

Levels of Scientific explanations:

- **Inference** – <u>a logical interpretation</u> based on prior knowledge or experience, e.g. *after a rainstorm, you see a bird's nest on the ground. You can infer that the rains washed away the nest.*

- **Hypothesis** – <u>a proposed scientific explanation</u>, which applies to a single or small number of events. This statement is testable and can be confirmed with experimentation or further observation, e.g. *not taking the full course of antibiotics, creates antibiotic resistant bacteria.*

- **Prediction** – An "if-then" statement that shows <u>what you expect to see</u> as a result of an experiment or observation, e.g. *"If fertilizer makes a plant grow faster, then seedlings planted with fertilizer will be taller than the ones planted without fertilizer."*

- **Theory** – <u>a widely accepted explanation</u> of natural phenomena. If a hypothesis is continually tested and is successful in predicting previously unexplained facts; the statement now becomes a theory, e.g. *The Theory of Natural Selection helps us understand why antibiotic-resistant bacteria develop from misuse of antibiotics.*

- **Law** – a statement of <u>what always occurs</u> under certain conditions. When a theory has proven to be invariable under all circumstances, then it may be a law, e.g. *"All living organisms arose in an evolutionary process."* However, note that laws do not explain "why" something happens.

LAB SKILLS IN CSEC BIOLOGY

B. <u>SAFETY IN THE LABORATORY – FOLLOW THE GUIDELINES!</u>

Accidents in the laboratory can be dangerous. They can be avoided by following these safety guidelines:

(A) BEFORE THE EXPERIMENT/ LAB: <u>read and try to understand ALL the instructions</u> given.

(B) ENTERING THE LAB:

1. Only enter with your teacher's permission.
2. Open all windows while you are in the laboratory, unless otherwise instructed.
3. Store bags, back-packs and purses in the designated areas away from the experiment setup.
4. Note the locations of exits and fire extinguishers; identify the safety showers and eye wash area.

(C) DURING THE LAB SESSION:

5. Do not eat, chew gum or drink beverages in the lab. Leave food and drink OUTSIDE the lab.
6. No playing or running in the lab. No smoking or applying make-up in the lab.
7. Follow all safety procedures during the experiment.
8. Do not leave an experiment, especially Bunsen burners unattended.
9. Do not pour unused chemicals back into stock reagent bottles – this is to avoid contamination.
10. Do not tamper with electrical mains and other fittings. NO charging of cell phones.
11. Wash your hands after every lab session.
12. Keep flammable substances e.g. ethanol away from lighted Bunsen or open flames.
13. Point test tubes away from other persons (and yourself) when heating; use a test tube holder.
14. Maintain a clean, uncluttered work area.
15. Do not dispose of solids – paper/ soil/ food materials down the sink. Use the dustbin.
16. Report all accidents, breakages and spillages immediately to your teacher or the lab personnel, no matter how minor! Sweep up broken glass, place in a "Broken Glass" box for proper disposal.
17. If any chemical enters your mouth or spills onto your skin, rinse/ flush with lots of water immediately.

(D) AT THE END OF THE LAB/ AFTER THE LAB SESSION:

18. Keep your area clean: wipe down counters and put away apparatus in the designated areas BEFORE leaving the lab area.
19. Return all stools to the designated area, away from the walk-way to avoid a trip hazard.

(E) CLOTHING/ PERSONAL PROTECTIVE REQUIREMENTS:

20. Protective clothing (lab coat/apron) is required when using chemical or biological agents.
21. Shoes must have closed toes and closed backs. No sandals, slippers or open-toed shoes allowed.
22. Long hair must be tied back when working with chemicals and open flames.
23. Gloves are required whenever there is the potential for contact with bio-hazardous materials. Disposable gloves must not be re-used especially if ripped, contaminated or dirty.

Student Agreement:

I have read and agree to follow these safety guidelines. I understand that the teacher/ lab personnel has the authority to remove me from the lab for failure to adhere to these guidelines. I also understand that the teacher may deduct marks for failure to obey them.

NAME (BLOCK LETTERS):_____**DATE:**_____

STUDENT'S SIGNATURE: _____**CLASS:** _____

LAB SKILLS IN CSEC BIOLOGY- Format

C. FORMAT OF A LAB: The parts of a lab report:

TITLE	A statement on what the lab is about.
AIM	• The purpose of the experiment. • Always begins with the words: • **to** determine…. or **to** investigate or **to** find out or **to** demonstrate… • And links the **manipulated (independent) and responding (dependent) variables.**
APPARATUS AND MATERIALS	• Apparatus are the different lab equipment/glassware supplied or used. • Materials are the chemicals, reagents or biological materials used. • USE A LIST – it is easier to read!
METHOD	• For most experiments, this is a set of steps written in **past tense using the passive voice** - *something* has <u>an action done to it</u>, e.g. "*The soil* <u>was weighed</u> on a scale.") • Number your steps so they follow a logical sequence. • For Plan and Design labs only, the method is a set of instructions in present tense.
Drawing/ Diagram	Include a 2D line drawing showing the set-up of the apparatus. * Except for drawing labs where more details are needed. (See drawing skill section.)

RESULTS	• This can be a number of the following:
	   • <u>A drawing(s)</u> of what you observed (drawing labs mainly)
	   • <u>A table</u> where you wrote down your data (responding variable)
	   • <u>A graph</u> that you plotted
	• NOTE – All tables must have a TITLE (above) and graphs must have a TITLE (below) describing what is contained within or shown.

DISCUSSION (USE PARAGRAPHS)	• This is like having a chat with someone about the experiment in formal language.
	• There are usually guiding questions that you should discuss.
	• Generally, points to discuss are:
	   • Background – theory about the topic (definitions, equations).
	   • The results – use data collected to describe trends AND then explain them.
	   • Particular steps in the method (precautions)/ when things went wrong.

LIMITATIONS	• Any **factor or variable which cannot be controlled** and might make **the results** less reliable. (See section on limitations explained)
PRECAUTIONS	• **Steps / methods the experimenter MUST take to ensure that the results are accurate**, e.g. specific use of chemicals, clean apparatus, reading at eye level, etc.
SOURCE OF ERROR	• An **error that may** <u>affect the results</u>. It may be due to equipment error(s) or experimenter inaccuracy, which could have been avoided if more careful.
CONCLUSION	• A clear statement that summarises the findings of the lab in ONE PARAGRAPH • It "answers" the aim by quoting some of your results.
REFLECTIONS	• Your **personal views** on <u>how the lab was useful to you</u>. Include, for example, what **you** thought but now know; how you can **use your understanding** and appreciation **in the future.**

LAB SKILLS IN CSEC BIOLOGY

D. LAB SKILLS IN CSEC BIOLOGY

1. MANIPULATION AND MEASURMENT (MM)

- **Manipulation** is the safe and correct **handling of equipment, apparatus, materials and living things**.
- You have to keep your **workplace** tidy and well organised. This includes apparatus and materials on the bench/table top, and the placements of **stools during the experiment and when you are done with the lab!**
- You should be able to manipulate glassware apparatus, chemicals, living organisms, the Bunsen burner, test tubes and microscopes.

- **Measurement** is a way to find the **quantity** of something, for example length and mass.
- Measuring instruments have a **scale (graduations)** marked on them – these must be read at the correct angle to get the correct value and **units** (cm/ ml or g).
- Instruments include – ruler, measuring cylinder, thermometer, stop clock and scale (balance).

SAMPLE CSEC MARK SCHEME FOR MM:

INVESTIGATING WATER HOLDING CAPACITY OF SAND AND CLAY		
Criteria - MM		**Marks**
General:	• Labelling samples and flasks	1
	• Correct setup of apparatus (measuring cylinder & funnel)	1
Scale:	• Zero the scale	1
	• Cleaning the platter of scale before weighing	1
	• Weighing the sand / Weighing the clay AND recording	2
Measuring cylinder:	• Measuring cylinder on a flat surface	1
	• Reading at eye level	1
	• Reading at the bottom of meniscus	1
	• No samples found in the m/cylinder after filtering	1
	TOTAL	**10**

The following pages describes and illustrates the most popular apparatus encountered in the biology laboratory with:

1. Simple 2D line drawings, and,
2. The steps to follow to successfully use the apparatus for its intended purpose, and to reduce experimenter errors.

LAB SKILLS IN CSEC BIOLOGY- MM

USE OF THE MEASURING CYLINDER

1. Ensure that cylinder is resting on a **flat, even surface.**
2. Read the **meniscus at eye level.** (see diagram on the right)
   Note: This requires stooping/ getting off the lab stool and bending.
3. Read the **bottom of the meniscus.**
4. Accurate **interpretation of scale** (cm^3or ml).
5. Ensure the measuring cylinder is empty before measuring, and ensure all liquid is emptied at the end.

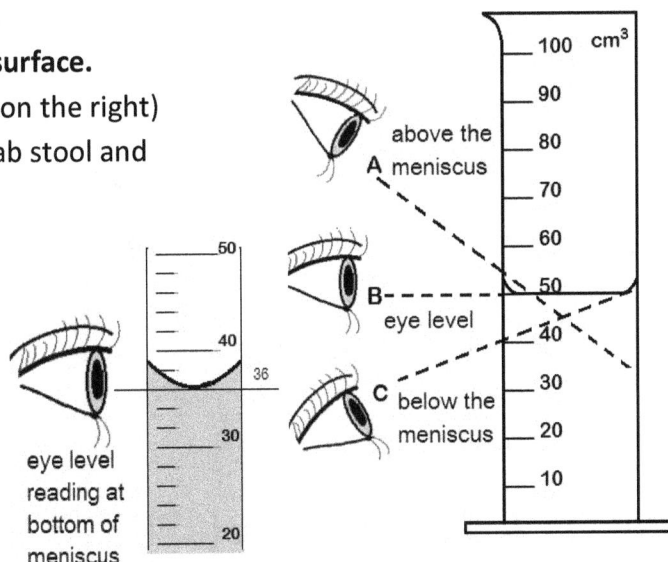

Meniscus – the curve in the upper surface of a liquid caused by surface tension.

USE OF THE THERMOMETER (in a liquid)

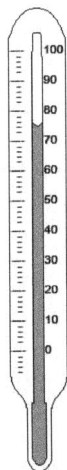

1. **Immerse bulb** completely in liquid.
2. Ensure the **bulb does not come into contact** with the container.
3. **Stir liquid** to ensure even distribution of heat.
4. Allow adequate immersion time for **equilibrium.**
5. Take **reading while the bulb is immersed.**
6. Take reading at **eye level.**
7. Handle carefully and store temporarily to **prevent breakage.**
8. Accurate interpretation of **scale** (correct reading in degrees Celsius): correct calculation of the number of smaller divisions (lines/ marks) between two numbers (longer lines/marks).

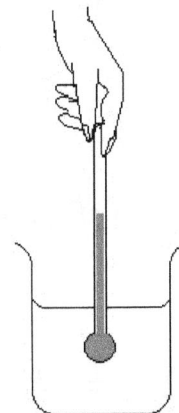

USE OF THE BUNSEN BURNER

Lighting
a. Keep air holes closed before lighting.
b. Strike/ light match <u>before</u> turning on the gas.

Adjusting the flame
1. Open air holes to obtain a non - luminous /clean blue flame.
2. Rotate the collar to obtain a clean, blue flame.
3. Control the size of the flame by adjusting the gas tap.

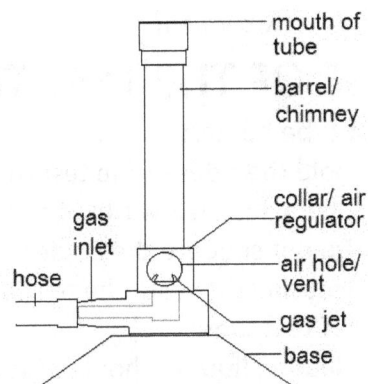

LAB SKILLS IN CSEC BIOLOGY

USE OF SYRINGE

1. Syringe is **clean and dry**.
2. **Plunger fully depressed** before filling.
3. Tip fully depressed before filling.
4. **No air bubbles** in the barrel with liquid.
5. **Top of rubber plunger** read at eye-level, upright.
6. Accurate interpretation of scale.

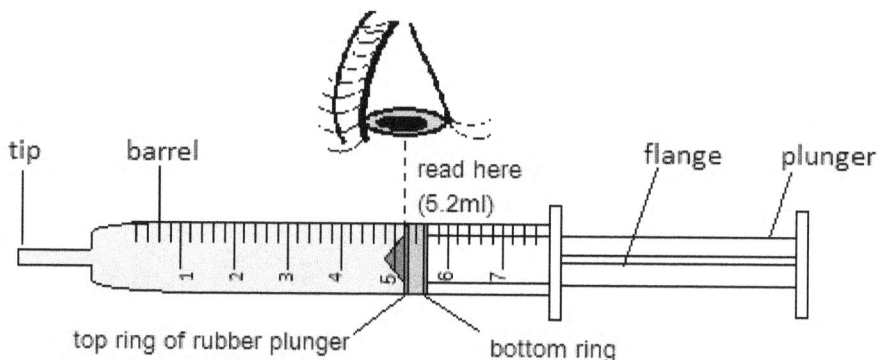

tip   barrel   read here (5.2ml)   flange   plunger

top ring of rubber plunger   bottom ring

USE OF A STOPCLOCK (SPRING WOUND)

1. Ensure the spring is fully wound up.
2. Start the clock at the same time as the event measured.
3. Ensure the clock is stopped as soon as the event stops.
4. Read the scale by looking directly at the marks.
5. Leave the clock working to run down the spring.

stop clock

USE OF A STOPWATCH (DIGITAL)

stop watch

1. Ensure time is zero centered.
2. Start stopwatch at beginning of reaction.
3. Stop stopwatch at end of reaction.
4. Accurately record time to the millisecond.
5. Proper handling– no swinging or throwing of equipment.

USE OF THE TEST TUBE HOLDER/ TEST TUBE

Test tube holder:
1. Hold the sides of the test tube – not the mouth.
2. Clamp just below rim of test tube.
3. Do not squeeze the holder (opening it) while moving a test tube (causing the test tube to slip out and fall).

Use of test tube:
4. **Heating liquids** - hold or face test tube <u>mouth facing away</u> from you and anyone nearby.

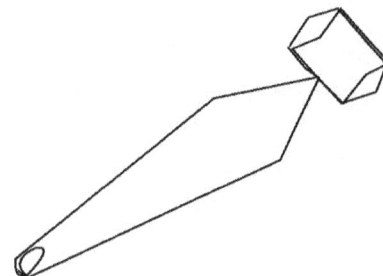

LAB SKILLS IN CSEC BIOLOGY- MM

HANDLING REAGENTS

- **Read label** before using reagent.
- Hold reagents/chemicals away from body.

1. Protection of labels: **Pour away** from the label (hand placed over the label).
2. Precautions to prevent contamination:
   a. Care of **stopper** - correct temporary storage upside down
   b. Stopper **replaced immediately** after use of the reagent.
   c. **Reagents already poured out must not be returned to stock reagent bottles.**
   d. If the reagent bottle is equipped with a dropper, use that dropper.
3. For accuracy with the dropper, add even sized drops.

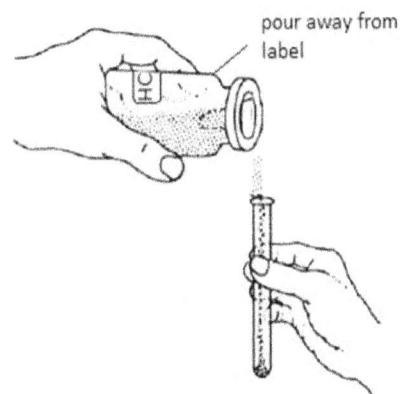

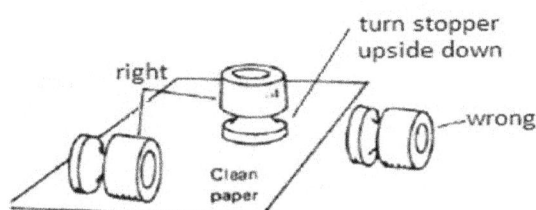

NO   YES

pour away from label

turn stopper upside down

right   wrong

Clean paper

CREATION OF A WATER BATH

test tube

beaker

tap water

gauze

tripod stand

flame

barrel

air vent

gas inlet

1. **ASSEMBLY OF WATER BATH**

   Bunsen burner, tripod stand, beaker, water

   a. Place away from books or paper.
   b. Position gauze accurately on tripod.
   c. Position tripod over Bunsen burner.

2. **BEAKER**

   a. Use appropriate sized beaker.
   b. 2/3 fill beaker with water.
   c. Carefully centre the gauze on the tripod.

3. **BUNSEN BURNER**

   a. Slide out Bunsen.
   b. Keep air vent hole closed (gas is off).
   c. Light match, turn on gas and light Bunsen.
   d. Adjust flame – turning the collar around the air vent so that there is a hole to the gas jet.
   e. Return Bunsen under the tripod stand.

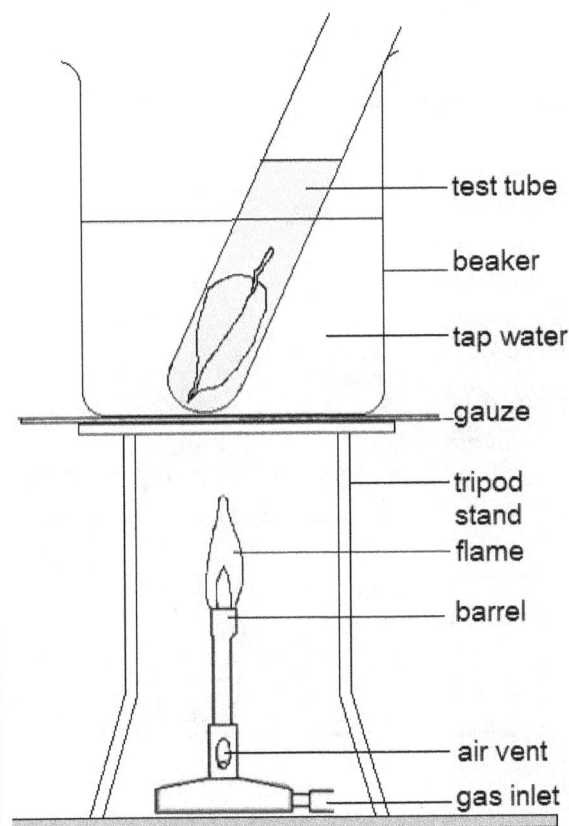

LAB SKILLS IN CSEC BIOLOGY

USE OF MICROSCOPE

1. Microscope must be held properly at base and arm.
2. Place microscope on a level surface.
3. Ensure objective lens is set to lowest power/ objective – x4.
4. Turn on light (only when needed).
5. Use coarse focus only with low power objective – x4.
6. Use fine focus at medium and high power objective lens – x10 and x40.
7. Look into the eye piece while focusing an object.
8. Lower stage to insert or remove a slide safely.
9. Replace microscope to X4 objective before putting away.
10. **TURN OFF LIGHTS WHEN NOT IN USE.**

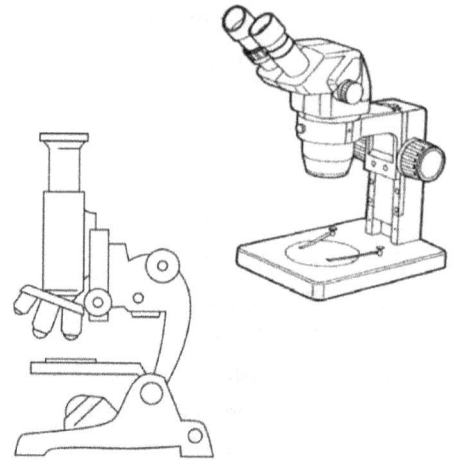

USE OF A DIGITAL ELECTRONIC SCALE

1. Plug in the scale and switch it on by pressing the "On/Off/Tare".
2. Set unit of measurement to grams or kilograms.
3. Ensure the platter is clean and free of dust.
4. Reset the zero value by pressing the Tare key.
5. Place the item to be measured on the platter in the middle.
6. Wait for the display reading to become stable.
7. Do not shake the table or press down on the platter with your weight or blow across the platter.
8. Record the reading displayed.
9. A beaker or petri dish can be placed on the scale, zeroed then liquid or salts added to the beaker for measurement.

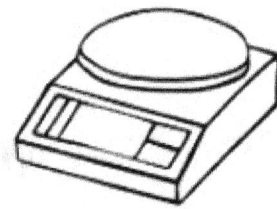

USE OF A RULER

1. Place ruler on a flat surface
2. Ensure zero value is at the end of specimen to be measured
3. Read the values directly over the point to be measured – avoid parallax errors.
4. Correct interpretation of scale – mm or cm.
5. If specimen is not straight or flat: use a string along its length and then measure the string with the ruler, starting at the zero on the left side.

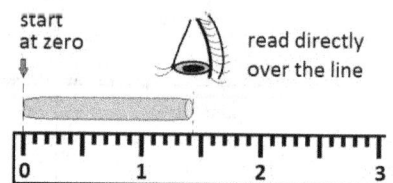

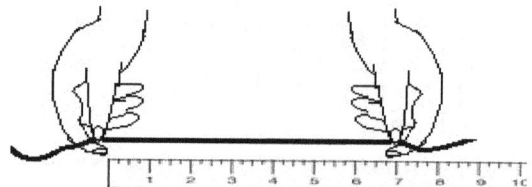

start at zero

read directly over the line

LAB SKILLS IN CSEC BIOLOGY- ORR

2. OBSERVING, RECORDING AND REPORTING (ORR)

A report should be clear and accurate so that someone who did not see the original observation or investigation can understand what took place. Students must have done the lab themselves and be confident in describing what they saw.

Keen Observations and Proper Recording = Great Reporting

- Observe using all your **senses**.
- Senses to use are **sight, touch** and **hearing**.
- **Smell** and **taste** are rarely used in the lab.
- Observe:
  - □ **volume** or
  - □ **mass** of substances,
  - □ **temperature** of a reaction
  - □ **length** of stems/roots,
  - □ **count** the number of organisms, and
  - □ **colour changes** in food tests,
  - □ **describe physical characteristics** – similarities and differences.

- Data can be **qualitative** (descriptions) or **quantitative** (numerical).
- Some observations must be done more than once and then **averaged**.
- **A good report** has accurate and precise observations and **records**.
- **Ways to record (or present) data** are:
  - – Tables
  - – Line graphs, Bar charts, Histograms
  - – Pie charts
  - – Diagrams or drawings
- Graphs or charts are made using data from the table.

Sample mark scheme for Observing, Recording and Reporting (ORR)

ORR GENERAL CRITERIA				Mark
O – Observations	S o c/d	• Significant changes noted • Original and final conditions compared • Control noted OR diagram		3
RTG – Recording		**TABLE**	**GRAPH**	3
	t u e	• Title – above, in capitals, underlined (1) • Column & row headings(with units) (1) • Enclosed and neat (1)	t • Title below, capitals, underlined (1) u • Both axes labelled with units (1) p • Accurate plots (1)	
R – Reporting	a g s r	• Aim in capital letters (1) • Acceptable language and expression – subject-verb agreement/ grammar (1) • Spelling correct throughout (1) • Reflection appropriate (1)		4
TOTAL				10

TABLES: Look at the sample table below – it shows the **parts of a table**.

The manipulated variable (independent variable) is usually the first column, followed by the responding variable (dependent variable).

Title → THE EFFECT OF EXERCISE ON THE BREATHING RATE OF FORM 1 STUDENTS.

Duration of exercise (mins)	Breathing rate (no. of breaths per minute)			
	Student 1	Student 2	Student 3	Average
0 – resting	50	55	57	54
1	60	65	65	63
2	69	75	77	74
3	80	85	89	85
4	80	86	89	85

Column (Vertical)
Units
Headings row(s)
Row horizontal
Borders

Table checklist:
- □ Title
- □ Columns
- □ Rows
- □ Headings in top row
- □ Units in header rows
- □ Values in cells
- □ No units (ml/cm) in cells; numbers only
- □ Borders all around

LAB SKILLS IN CSEC BIOLOGY

LINE GRAPHS:

A line graph shows the relationship between two numerical variables. The experimenter changes the values of one (referred to as independent/manipulated variable) - such as time or temperature.

1. **Use a PENCIL and ruler** to draw on a graph page or grid. The graph should be large – use at least half the page.
2. The **horizontal** axis - **X axis is the independent variable** – it is what the experimenter changes or can control – manipulated variable, e.g. time in hours.
   The **vertical** axis – **Y axis is the dependent variable** – what was measured/recorded – responding variable.
3. Choose the scales: evenly divide each axis by making small lines. They do not need to start at zero.
   Write your scale in a corner of the graph page (e.g. x-axis: 2cm represents 0.25hr and y-axis: 1cm represents 10mg)
4. **Label the axes** and place their units in brackets.
5. **Plot points** with either a dot in a circle O or a neat **x** or **+**.
6. **Join points** one after the other using a ruler. Do not extend the line beyond the last point.

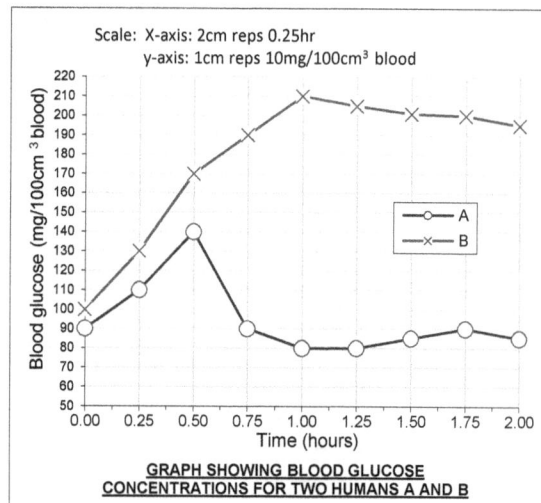

GRAPH SHOWING BLOOD GLUCOSE CONCENTRATIONS FOR TWO HUMANS A AND B

7. **Add a TITLE** below the graph. The title must be descriptive and meaningful. e.g. "EFFECT OF TEMPERATURE ON OXYGEN PRODUCTION IN LEAVES".
8. **Use a key** if two or more sets of results are plotted on the same axes. DO NOT divide the graph paper in half to draw separate plots with separate axes.

BAR CHARTS:

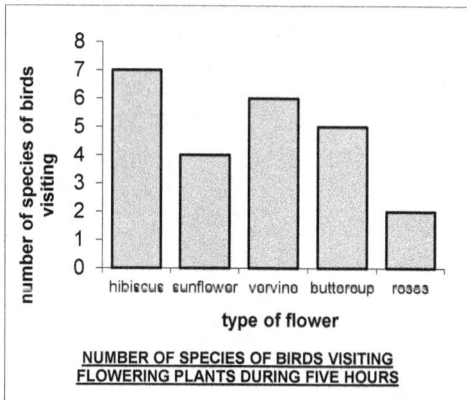

NUMBER OF SPECIES OF BIRDS VISITING FLOWERING PLANTS DURING FIVE HOURS

Used when only **one variable is qualitative (descriptive)** (independent variable on the x-axis) and **the other is numerical** (dependent variable on the y-axis).

1. Draw EVERYTHING IN PENCIL on graph paper. Use a ruler.
2. There are 2 axes (include names and units for each).
   - X axis – horizontal axis – independent or manipulated variable e.g. type of flower.
   - Y axis – vertical axis – dependent or responding variable e.g. Number of species of birds visiting.
3. There should be a DESCRIPTIVE TITLE below the chart.

HISTOGRAMS (FREQUENCY DISTRIBUTION):

Histograms are a special kind of bar chart. They are used when both variables (manipulated and responding) are numerical and one of the variables is continuous and can be grouped into sets.

- Use graph paper, a ruler and a PENCIL.
- Group the manipulated variable into **ranges** – X units apart.
- **Bars MUST TOUCH** each other (data is continuous).
- **Label each axis**, put units in brackets, and include a scale on the graph.
- Give a **DESCRIPTIVE TITLE** below the graph.

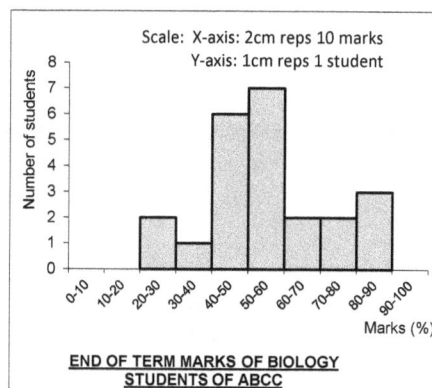

END OF TERM MARKS OF BIOLOGY STUDENTS OF ABCC

LAB SKILLS IN CSEC BIOLOGY- ORR

REFLECTIVE WRITING – FOR BIOLOGY/SCIENCE

A reflection in science is supposed to help you **understand the relevance** between the experiment and real life (yourself, society or the environment). It should indicate what **impact the knowledge gained** from the experiment **had on you**.

You need to practise writing reflections for each lab since your project (a plan & design lab you make in Form 4, Term 3, then carry out in Form 5, Term 1) will require you to write a reflection to gain valuable SBA marks. Reflections for Biology lab experiments can be kept short.

The main parts in your reflection will be

1. **Description** – What happened or what was examined? – This is already done in the discussion.

2. **Interpretation** - What was most important/interesting/useful/ relevant about the topic or idea? And, how can it be explained?

3. **Outcome** – What have you learnt from this and what does that mean for your future?

To give you some ideas on how to word your sentences for your reflections use the guide below.

Remember, for the most part you are only writing a few sentences under **interpretation and outcome**

2. **Interpretation** (probably the most important bit)			
For me, the (most)	meaningful / significant / important / relevant / useful	aspect(s) / element(s) / experience(s) / issue(s) / idea(s)	was (were)…
		learning	arose from… / happened when… / resulted from…
	Previously, / At the time, / At first / Initially, / Subsequently, / Later,	I	thought (did not think)… / felt (did not feel)… / knew (did not know)… / noticed (did not notice)… / questioned (did not question)… / realised (did not realise)…
[Alternatively,] / [Equally,]		This	might be / is perhaps / could be / is probably	because of… / due to… / explained by… / related to…

3. Outcome

Having	read… experienced… applied… discussed… analysed… learned…	I now	feel… think… realise… wonder… question… know…	

[Additionally,]
[Furthermore,]
[Most importantly,]
[In addition,]   I have learned that…

I have

However, I have not [sufficiently]   significantly
slightly   developed
improved   my skills in…
my understanding of…
my knowledge of…
my ability to…

This means that…
This makes me feel…

This knowledge This understanding This skill	is could be will be	essential important useful	to me as a learner [because…] to me as a practitioner [because…]

Because I   did not…
have not yet…
am not yet certain about…
am not yet confident about…
do not yet know…
do not yet understand…   I will now need to….

As a next step, I need to…

Example of a reflection:

"In this experiment to see which source of sugar that ants prefer, I learned a few lessons. Firstly, it is very difficult to catch ants, and ants can bite. The results also showed that ants preferred natural sugars compared to artificial sweeteners. This information is useful to me as it is an indication that if ants choose a more natural food, then I should also choose a more natural food. This is important in proper diet and making wise food choices."

LAB SKILLS IN CSEC BIOLOGY- AI

3.   ANALYSIS AND INTERPRETATION (AI)

When you have carried out your practical, collected and recorded your data, you have **to show you understand the scientific principles and to explain them.**

You should be able to:

- Read values – off the tables/ graphs
- Identify patterns and relationships – on the graphs
- Carry out calculations
- Evaluate what you have found out – why those results are important
- Make predictions
- Draw conclusions – **sum up your findings** /say whether **your results** (dis)agree with your hypothesis.

In Biology, the skill of AI usually looks at the **discussion and conclusion** to award the marks.

Three main aspects of the discussion for AI skill are:
1. **Biological concepts** – the background or theory of the topic (check your notes).
2. **Explaining steps** in the method – including precautions.
3. **Interpreting and evaluating your results:**
   - what the data showed (describing) and why (usually linked to biological concepts);
   - sources of error or inaccuracies and how to deal with them/ or how they were minimized.

The **conclusion** also contributes to this skill: it is a logical **summary statement** about the findings.

ANALYSIS:
You may have to do some calculations or plotting first. Then look at your tables, line graphs or histograms for:

- **Trends/Patterns** – data that are repeated in a similar way so a prediction can be made. e.g. if a line is increasing, then if the line is continued, a further increase is expected.

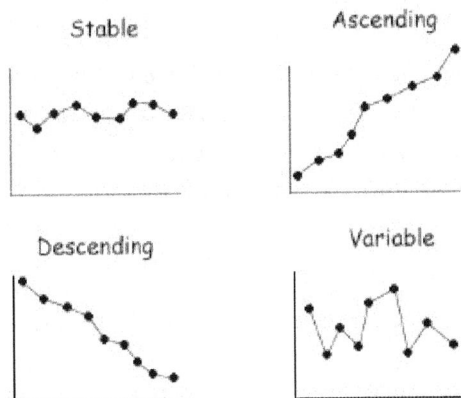

Stable   Ascending

Descending   Variable

- **Relationship** - how variables (manipulated and responding) are related. e.g. the line drawn on a graph shows if values increase together, or decrease together (based on the slope)

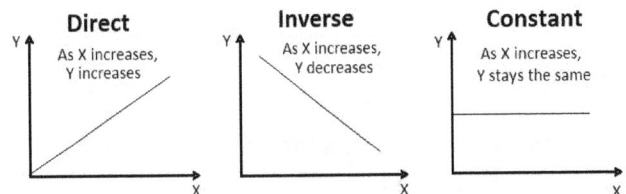

Direct	Inverse	Constant
As X increases, Y increases	As X increases, Y decreases	As X increases, Y stays the same

Where: X is the manipulated variable and Y is the responding variable

INTERPRETATION:
After finding relationships and patterns, interpretation usually involves
- **Explaining** – a reason for your pattern or relationship. Compare with the control.
- **Linking** – to scientific facts or the theory.
- **Predicting** – another expectation or pattern if some other variable were changed or added.
- **Identifying** – sources of error/ limitations.
- **Conclusion** – you final statement on the experiment, a summary.

LAB SKILLS IN CSEC BIOLOGY

Interpreting data on graphs: Example 1

The table below represents the results of an experiment to determine the effect of pH on the rate of starch breakdown by an enzyme.

TABLE SHOWING AMOUNT OF PRODUCT MADE BY ENZYME PER UNIT TIME AT VARIOUS pHs

pH	AMT. PRODUCT PER UNIT TIME/mg
2	0
4	0
6	65
8	20
10	1
12	0
14	0

Questions

a) Draw a graph to represent the results.

b) Describe what the graph shows and explain why it is shaped this way.

c) Where in the human digestive system is this enzyme found?

d) Suggest what happens if there is not enough of this enzyme or any other enzyme available.

(a) Plotting the graph:

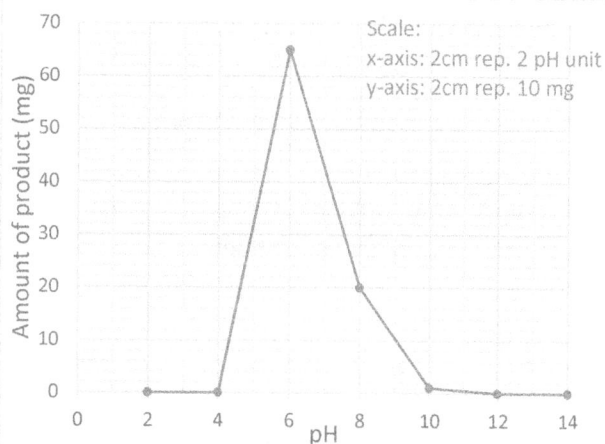

Scale:
x-axis: 2cm rep. 2 pH unit
y-axis: 2cm rep. 10 mg

GRAPH SHOWING THE AMOUNT OF PRODUCT PRODUCED BY ENZYME AT VARIOUS pH.

Graph checklist

☐ Choose the right axis for each variable
☐ Label the axes and put units
☐ Use proper scale for data (use most of graph page)
☐ Plot points – use symbol - ☉ or X or +
☐ Join points with straight lines
☐ Make a key (where 2 data sets plot on same axes)
☐ Title – appropriate and descriptive

(b) Interpreting the Graph

Describe: - State precisely the effect that changes in one variable (INDEPENDENT variable) have on the other (DEPENDENT variable).

In the above example, as the pH increases from 0 to 2, there is no activity of the enzyme but as it further increases from 4 to 6 the quantity of product formed per unit time increases to a maximum of 65 mg. However, above pH 6 the rate gradually decreases and returns to zero after pH 10.

Explain – you must give a REASON for the effect of changes in one variable on the other. You usually have to use previous knowledge (theory) to explain the effect shown on the graph.

In this example, the reason for the shape of the graph is that enzymes can only work within narrow pH ranges, hence there is enzyme activity only between pH 4 - 8. The optimum pH in this case is pH 6.

Parts (c) and (d) ask you questions on information understood from the graph.

(c) *If this enzyme was found in a human, it can be in either the mouth or the small intestine. Both these places are neutral to slightly alkaline, the optimum for this enzyme.*

(d) *If no amylase is present in the body then there will be the inability to ingest starches. Furthermore, there will be a lack of energy as amylase will not be able to break down starch into maltose sugar. Glucose is the main source of energy especially for the brain cells to respire.*

Interpreting data on graphs: Example 2.

The basic principles of aquaponics involves raising fish and prawns in tanks alongside hydroponics – cultivating plants in water. The excretions from the animals are toxic to them and can accumulate in the ponds. But when fed through a hydroponic system, nitrifying bacteria can form nitrates for plants to use in their growth.

The average growth of two sets of seedlings were recorded over a month, using different sources of water as in the table. Answer the following questions based on the information in the table.

(a) Draw a graph using a **single** sheet of graph paper to represent these results.

(b) Compare what is shown by the two curves.

(c) How could you account for the difference in the heights of the seedlings?

AVERAGE HEIGHT OF SEEDLINGS GROWN IN DISTILLED WATER AND POND WATER FOR A MONTH.

TIME/ DAYS	AVG. HEIGHT OF SEEDLINGS (CM)	
	Distilled water	Pond water
0	7	7
2	8	9
4	10	12
6	12	15
8	15	20
10	16	22
12	16	24
14	22	28
16	23	30
18	24	32
20	24	36
22	25	40
24	26	46
26	26	48
28	26	52
30	26	58

Answers:

(a) Use the checklist in example 1 above to draw the graph.

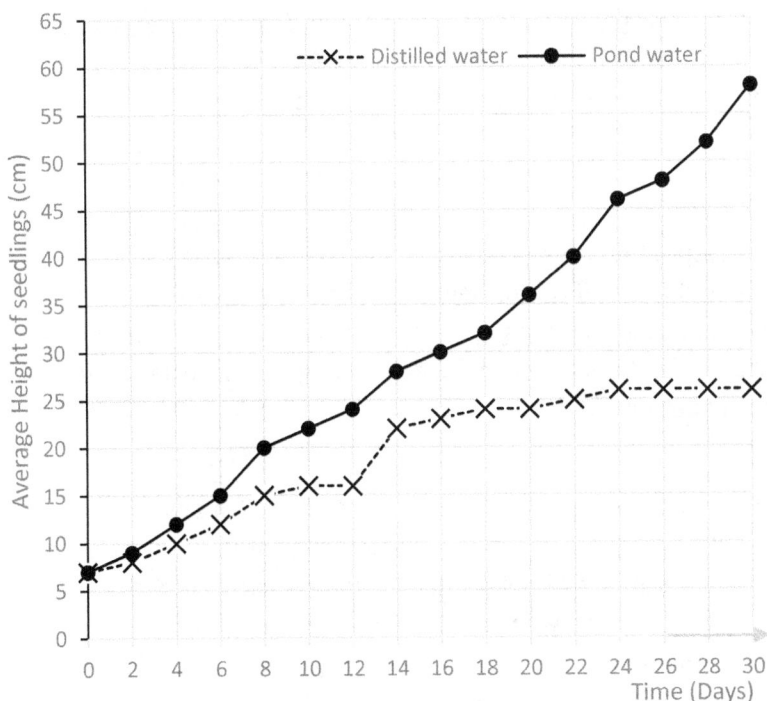

GRAPH SHOWING AVERAGE HEIGHT OF SEEDLINGS GROWN IN DISTILLED WATER AND POND WATER

(b) The graph shows that in general, plants grown in distilled water did not grow as tall as those grown in pond water over 1 month. During the first 8 days, the growth rates were almost the same, but from day 8 onwards, the seedling in pond water kept increasing steadily in height up to 58cm. Whereas growth became almost constant for the seedlings in distilled water, with a very slight increase from 22cm up to a constant 26cm in height from day 14 till day 30.

(c) After 30 days, the seedlings in pond water grew taller than those grown in distilled water because more nitrates were made available for the plant tissues to grow. Plants use nitrates as a source of plant proteins, needed for development of new healthy tissues. The plants grown in distilled water was not able to get any additional nutrients to increase stem length.

LAB SKILLS IN CSEC BIOLOGY

Sample mark scheme for Analysis and Interpretation – AI

AI GENERAL CRITERIA				SPECIFIC CRITERIA (to be defined…)	MAX MARK
B- Background	d	•	Define key related terms (1)		2
	s	•	Statement of relevant theory (1)		
E -Explanation	t	-	Trends and patterns identified (1)		4
	c	-	Compare actual results with expected results (1)		
	u	-	Use data to support explanations (1)		
	m	-	Modification/ improvement to existing method (1)		
LSP -Limitations Sources of error/ Precautions		-	Identify at least 2 limitations (with explanations)		2
		-	Identify at least 2 precautions/ sources of error (with explanations)		
		-	**any 2 of the above**		
C- Conclusion	s	-	Statement		2
	r	-	Related to aim		
TOTAL					10

Why use a 'control'?

When testing a theory/ hypothesis, it is important to have a control. The control acts as **a neutral experiment to compare other results;** sometimes **the control is the normal situation**. When comparing a control situation to the experiment situation, it is important to ensure that only one factor is being varied. Sometimes it is not possible to have a control, such as when using humans as the test subject.

Examples of controls:

1. In enzyme type experiments, the control is <u>usually</u> a test tube that contains <u>distilled water,</u> while all the other tubes contain enzymes. If the results of the tube containing distilled water (a neutral substance) and the tube containing the enzyme are different, than it can be concluded that the difference was due to the enzyme.
2. Other examples of controls can be <u>dead beans (non-germinating/boiled beans)</u> – in the case of investigating if germination releases heat.
3. <u>Boiled/ killed yeast</u> when investigating respiration in yeast.
4. <u>A leaf that receives no sunlight</u> when investigating if light is needed for photosynthesis.
5. When testing a new pharmaceutical drug, one group of <u>patients is given a placebo (a substance with no effects)</u> whereas the other group of patients is given the actual drug. This is to see if by just taking a tablet, the patients feel/ get better.

Controls are not Constants!

Constants are variables (conditions) that remain the same throughout the experiment. For any experiment, there is usually more than one condition held constant. For example the constant variables in a seedling growth experiment will be the mass of soil, size of pots, volume of water used every day, exposure to sunlight and number of seeds in each treatment, whereas the control in an experiment depends on what variable was manipulated. For example, if the aim was to investigate <u>the effect of type of water on growth rate </u>(see graph example 2 on previous page), then watering with distilled water will be the control experiment.

LAB SKILLS IN CSEC BIOLOGY- AI

Limitations:

Limitations are **variables which you may not be able to control or eliminate**. These can **affect the results** of the experiment in some way. Limitations must be recognised and accounted for in your discussion of the experiment especially when they cannot be corrected for.

Examples of limitations and precautions:

NO.	TOPIC and AIM/ METHOD/ SCENARIO	LIMITATIONS AND PRECAUTIONS
1.	Osmosis - Using potato cylinders in different concentrations of sucrose solutions and measuring length before and after immersion	Limitations • All cylinders may not come from the same tuber so they may each have different concentrations of cytoplasm in their cells. • The age of the tubers (recently formed vs starting to grow again) will affect the concentration of sugars/ starch stored within. Precautions • Totally immerse cylinders in the solutions. • Keep the experiment at constant temperature. • Keep each cylinder in the solution for the same length of time. • Potato cylinders should be the same dimensions. • Volume of water/ solutions should be the same and accurately measured to ensure comparisons are fair.
2.	Ecological sampling - Quadrat sampling and Line transect sampling	Limitations • Tall plants cannot be sampled with small quadrats. • Quadrats are point samples and do not show zonation or succession (change in type of plants) along a line. • Transects alone cannot determine species frequency, density or cover; quadrats are needed for those measures of abundance. Precautions • Throw quadrats randomly to obtain fair results. • Repeat the throws more than once to increase accuracy of results.
3.	Photosynthesis - To see if a plant photosynthesizes faster in red vs. white light using bulbs next to them	Limitations: Heat produced by the light bulb can increase temperature (another variable that affects rate of photosynthesis) thus speeding up photosynthesis. One solution may be to use a heat filter, e.g. a beaker of water.
4.	Photosynthesis - To determine if colour intensity of extracted chlorophyll depends on where a leaf was growing – in the shade or full sunlight	Limitations • Colour intensity is subjective (non-numerical)/ different based on persons deciding the intensity. • Colour intensity may not accurately represent the chlorophyll content (assumption that only chlorophyll pigments are extracted). • Some chlorophyll could be lost during grinding of leaves. • Waxy leaf surfaces could affect chlorophyll extraction since using too much ethanol dilutes the colour obtained.
5.	The eye/ Reflex actions - The effect of light intensity (distance between light and eye) on pupil size	Limitations • Eyes may get tired so reactions slow down. • Fluctuations in electricity/ light intensity shone in the eyes. • Unable to measure the actual size of the pupil accurately.

LAB SKILLS IN CSEC BIOLOGY

Precautions and Sources of Error:

Precautions are **steps in the method for the experimenter to ensure the results are accurate**. If precautions are not taken, then they may become potential sources of error, i.e. reasons why the results are not what was expected. Examples of precautions and sources of error:

1. In an experiment to determine the effect of temperature on amylase enzyme activity, the following precautions should be taken in the method:

   i. Use clean dropper / test tubes.
   ii. Use equal volumes of reagents (enzyme and substrate) at each set-up.
   iii. Keep mixtures in respective water bath for 15 minutes.
   iv. Remove sample precisely after 15 minutes.
   v. Mix the reagents thoroughly.
   vi. Keep water bath at constant temperature.
   vii. Use thermometer properly by submerging the bulb in mixture.

If the precautions identified were not followed, then they may become sources of error:

   i. Using contaminated apparatus may result in inaccuracies, starch does not breakdown.
   ii. Different volumes of reagents could mean the experiment was not fair.
   iii. Not mixing the reagents makes some of the substrate unavailable for enzymes to break it down.
   iv. The end point may be missed and critical colour changes not observed if left for less than 15 minutes.
   v. Fluctuating temperatures could have resulted in lower rates than expected.
   vi. Using/ reading thermometer inaccurately due to parallax error so reaction tubes were at a lower temperature than needed, making the rate of enzyme activity also lower than expected.

2. When testing if small organisms respire or produce carbon dioxide.

   Precautions taken should be to:
   i. Ensure the organisms are alive
   ii. Ensure there are no other microorganisms in the chamber
   iii. Maintain constant room temperature
   iv. Ensure the set-up is air-tight.

   If the limewater does not turn cloudy at all, then the source of error could be that:
   i. The organisms have died.
   ii. There was a leak in the set-up.
   iii. The temperature was too cold for the organisms to respire.

Assumptions:

Assumptions are **statements that are accepted as true**. Assumptions should be known where possible before the start of the experiment (a plan and design) and identified in the discussion section.

Examples of assumptions:

1. Transpiration: experiment using a potometer (measures the volume of water stems absorb). It is assumed that all the water that a plant takes in is lost through transpiration (through the stomata of the leaves). But plants actually use some water for photosynthesis and to maintain cell turgidity, even though most of the water is lost by transpiration.

2. Photosynthesis – testing a leaf for starch:
   Although the majority of the leaves store starch, not all leaves store that particular carbohydrate. Some leaves, e.g. onion and chive, contain glucose.

4.   DRAWING (DR)

Scientific drawings are an important part of the science of biology and all biologists must be able to produce good quality scientific drawings regardless of your artistic ability.

Purpose of drawings for biology
- **Record** a specimen observed.
- Help you to **remember the specimen**, especially for exams.
- **Identify important features**.
- **Relate structures to functions**.

Additional skills gained by drawing
- You learn to **pay attention** to detail.
- You **concentrate** on a task.
- You can **re-create a specimen**.
- Your **Memory is improved**.

WRONG

All drawings done for SBAs must adhere to standard rules of scientific illustration.
The following are some guidelines for to use when illustrating specimens:

A)   DRAWING
1.   Look at the specimen carefully, note the significant features to include in the drawing.
2.   **DRAW ONLY WHAT YOU SEE,** NOT what you think you should see.
3.   All drawings must be done in pencil ONLY.
4.   Drawings must be **large** and **clear** so that features can be easily distinguished.
5.   Always use **distinct, single lines** when drawing. DO NOT SKETCH/ SHADE.

BETTER

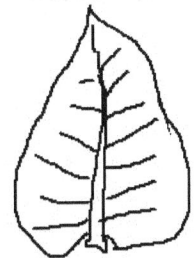

B)   LABELLING
6.   Use a ruler to make straight label lines; touch the part labelled.
7.   Do not cross labels OR put arrow heads to label lines.
8.   Put all labels to the right of the drawing and justify (line up one under the other).
9.   Write labels IN CAPITALS OR in non-cursive text – that is – the letters are not joined.
10.   Include at least one **annotation** this may be **functions** or **descriptions** next to labels.

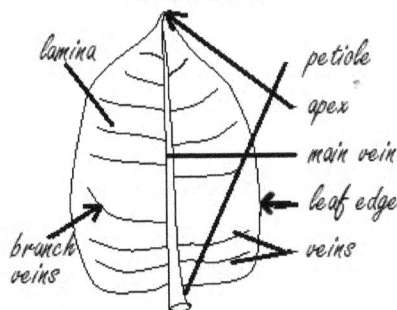

Drawing of Ixora leaf

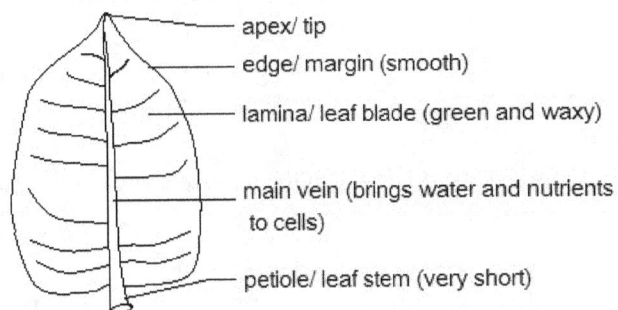

DRAWING SHOWING IXORA LEAF EXTERNAL VIEW (MAG. X1)

C)   TITLE
11.   The title must be written **below** the drawing in **CAPITAL LETTERS,** and **UNDERLINED!**
12.   Indicate the **section (cross section, longitudinal section)** or **view (external view, internal view)**.
13.   Place the **magnification** of the drawing **in brackets** (Mag. x 2.0). Report to 1 decimal place (1d.p.)

$$\textit{Magnification of drawing} = \frac{\textit{length of the drawing (cm)}}{\textit{length of the specimen (cm)}}$$

Note: magnification has **no units** as they cancel out.

LAB SKILLS IN CSEC BIOLOGY

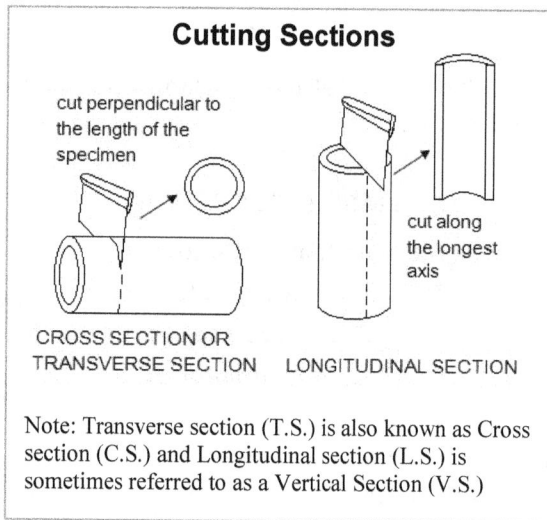

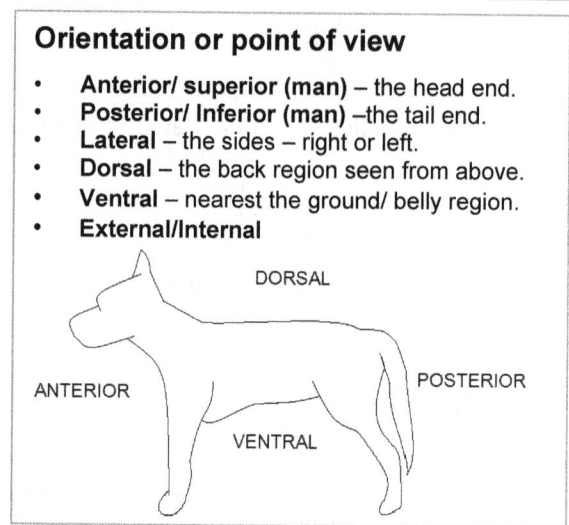

Cutting Sections

cut perpendicular to the length of the specimen

cut along the longest axis

CROSS SECTION OR TRANSVERSE SECTION

LONGITUDINAL SECTION

Note: Transverse section (T.S.) is also known as Cross section (C.S.) and Longitudinal section (L.S.) is sometimes referred to as a Vertical Section (V.S.)

Orientation or point of view

- **Anterior/ superior (man)** – the head end.
- **Posterior/ Inferior (man)** –the tail end.
- **Lateral** – the sides – right or left.
- **Dorsal** – the back region seen from above.
- **Ventral** – nearest the ground/ belly region.
- **External/Internal**

DORSAL

ANTERIOR

POSTERIOR

VENTRAL

Sample of drawing mark scheme - DR

CRITERIA (DRAWING – DR)				MAX. MARKS
C- CLARITY	l	-	Large drawing (½ page or larger) (1)	3
	c	-	Clean, smooth, thin, continuous lines (1)	
	s	-	No shading or unnecessary details (1)	
A- ACCURACY	s	-	Looks like specimen (1)	2
	p	-	Reasonable proportions – length vs width (1)	
L-LABELLING and LABEL LINES	p	-	Straight, no arrow head, parallel (1)	3
	a	-	Accurate label, lowercase, letters not joined (1)	
		-	Annotations – at least 3 (1)	
	j	-	Justified labels (start at some point)	
	m	-	Magnification calculation correct (1)	1
	t	-	Title – at bottom, in CAPITALS, underlined	1
		-	View of specimen stated	
TOTAL				**10**

Drawing example: - the drawing below shows many different drawing errors. Identify them!

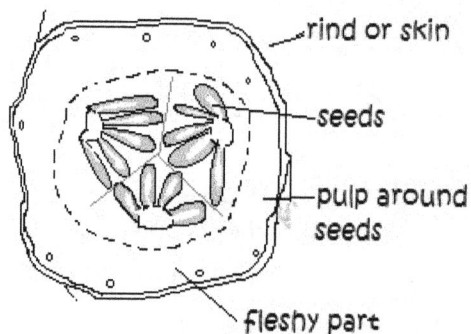

rind or skin

seeds

pulp around seeds

fleshy part

CUCMBER IN SECTION

Errors in the drawing:
1. Drawing is neither clean nor has continuous lines.
2. There is shading and lines of varying thickness.
3. Label lines are not touching the figure (rind).
4. A ruler was not used to draw label lines which must be parallel.
5. There are no annotations for the labels.
6. Wrong title: "cucumber" spelt incorrectly and the "section" must say "cross section". The title needs to be underlined.
7. Calculation of magnification not stated.

LAB SKILLS IN CSEC BIOLOGY- PD

5.   PLAN AND DESIGN (PD)

The layout or format for a plan and design lab is a little different from what you are accustomed to. Here is a brief overview. The most difficult part for students is usually formulating and writing the hypothesis, but that will be explained later in this section.

Format and sample mark scheme for Plan and Design Labs (PD)

	CRITERIA FOR PD and EXAMPLES	Mark
Observation/ Situation	The student can observe a phenomenon or regular occurrence and state it. OR The teacher can provide an observation to consider planning an experiment. e.g. Hummingbirds are often seen feeding on red coloured flowers in the garden and most hummingbird feeders are red.	
Hypothesis	A plausible/ logical statement (**1mk**) that is testable (**1mk**) (See below) e.g. More hummingbirds visit red feeders than any other colour.	**2**
Aim	The Aim is related to the hypothesis (**1mk**), but it must be stated as an aim and it must also take into consideration the method. (Variables are identifiable.) e.g. To count and compare the number of hummingbirds visiting red, yellow and green feeders, OR: To investigate the effect of feeder colours on number of visits by hummingbirds.	**1**
Apparatus and Materials	List of appropriate apparatus & materials to be used (**1 mark**) To create this list, you may need to outline the procedure first, thinking through all parts and questions (see "Planning the method for a PD lab" below).	**1**
Procedure (Method)	Written as INSTRUCTIONS or in present tense (**1mk**). (You have not yet done the experiment!) -logical steps/ sequences in point form (**1mk**) -repetition of procedures for accuracy when collecting results (**1mk**) (see "Planning the method for a PD lab" below)	**3**
Variables	Relist the following variables if not already listed in your method: 1. Manipulated – what **one feature** you changed, e.g. colour of the feeder 2. Responding – what measurements were recorded, e.g. number of visits 3. Constant – all the features kept the same between each experiment set-up, to make the experiment a fair one.	
Control	If not already identified in your method, the control may be re-stated here. Remember the control is one experimental set-up where the manipulated variable was absent/ not conducted.	**1**
Expected results & Interpretation (Explanation of expected results)	- Expected results: - Must be **clearly stated in words, a paragraph explaining what is expected and the reasons for it**. You may need to do research online. - Additionally, include a blank table to collect results with headings and a title. Or, if a graph is expected, sketch the axes and describe the trends expected. - The interpretation must indicate how to analyse the data/ patterns expected. - In addition state what the expected occurrence is if you are **to accept the hypothesis**, AND, what to do next **if you reject your hypothesis** (and why that may occur).	**1**
Limitations	Should include shortcomings of the design which may become a source of error, e.g. the same bird may constantly be at one feeder, not giving others a chance; it is very difficult to watch all the feeders at the same time – need more observers.	**1**
	TOTAL	**10**

LAB SKILLS IN CSEC BIOLOGY

WHAT IS A HYPOTHESIS?

- A hypothesis is **a tentative statement** that gives a **suggested reason** for an observation and can be **tested in the lab, i.e. is testable**.
  - ○ Tentative – since it can be revised/ improved later on.
  - ○ Suggested reason – if the hypothesis is proven true, it may be used to explain similar observations/ occurrences.
  - ○ Tested in the lab/testable – when stated correctly, the independent and dependent variables are highlighted. Furthermore, by making observations, and controlling conditions the statement can be accepted/ proven (correct) or the statement can be rejected (wrong).
- Three features of a good hypothesis statement are that:
  1. It relates to the observation made directly
  2. It can be experimented on in the lab
  3. Only one condition or variable is changed/ manipulated.
- A hypothesis is NOT A GUESS. A hypothesis is a reasoned explanation based on scientific knowledge.

Developing a hypothesis

Example 1: Thinking it through step by step

Observation: In general, male guppies (*Poecilia reticulatae*) are colourful in order to attract the female and they reproduce quickly. In the wild, guppies are found in small streams in the Trinidadian Northern range. These streams are punctuated by waterfalls, but guppies and other fish, such as the Rivulus fish (*Rivulus hartii*) which sometimes eat guppies, are found living above and below the waterfall. During the rainy season, guppies swim upstream when rivers overflow. A voracious predator of the guppy is the pike cichlid (*Crenicichla alta*), that is only found in pools below the lower waterfall.

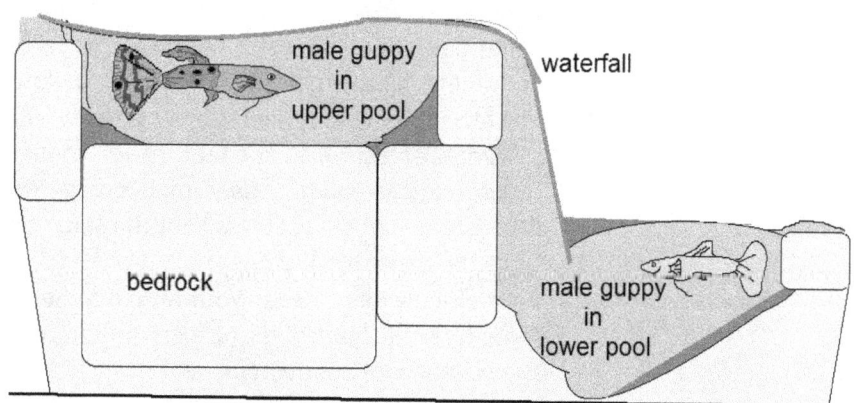

It was observed that the male guppies in pools below waterfalls are drab coloured and smaller, compared to the male guppies above the waterfalls; which are larger, colourful and attractive to females.

Step 1. – Identify possible reasons or explanations:

Each one of these can be used as a possible hypothesis and can be tested:
- The female guppies in higher pools are pickier.
- Higher pools are larger than those below.
- There are more guppy predators in the pools below waterfalls.
- Only drab coloured guppies wash down into lower pools.
- The larger guppies are stronger and can swim upstream.

Step 2. – Choose an explanation and do more research:

a. Research similar studies and compare their findings with your observations.
b. Link the facts with the observations.
c. Identify possible questions – Do other river fish have similar colour/ size differences? What other fish (possible predators) live in those two pool environments? What effect do predators have on male guppy colours? Are the two populations of guppies related?

Step 3. – Formulate a testable hypothesis from the explanation:

A testable hypothesis is a clear statement, making a prediction based on the explanation.

Possible hypotheses:

- Female guppies prefer brightly coloured male guppies for reproduction.
- Larger pools support larger sized male guppy fish.
- Being more colourful and larger increases the chances of predation in male guppies.

Example 2. - A hypothesis can be started off written as an "if….. then….(because….)" statement.

Step 1: Based on the observations, the variables are identified and an effect predicted:

If   flower colour is related to   preference in feeding of hummingbirds   then hummingbirds prefer red flowers.

(independent variable)   *(dependent variable)*   *(predict the effect)*

Step 2: Rework the sentence:

Providing hummingbirds with <u>feeders of varying colours</u>   more hummingbirds visit the red feeders.
(describe changes in the independent variable)   then   *(describe the effect on the dependent variable)*

Step 3. – state it as a hypothesis:

More hummingbirds visit red feeders   presented with a variety of coloured feeders.
(dependent variable effect predicted)   when   *(describe the change in the independent variable).*

LAB SKILLS IN CSEC BIOLOGY

Planning the method for a PD lab:
Example 2 – Hypothesis – More hummingbirds visit red feeders than any other colour.

Think carefully about what you want to find out. Consider these questions to formulate the method:

1. Observations and Measurements (recording results) :

☐ What measurements must be taken?	-	Counting the number of birds visiting a feeder?
☐ Should other things be measured as well?	-	Time of day/ weather?
☐ How will they be measured?	-	Using a camera, or sitting at a bird feeder? Should I time a bird; what if it's the same bird visiting more than once?

2. Time/ duration:

☐ How long will the investigation take?	-	A few hours, a week, a month? E.g. 5 days.
☐ How often must measurements be taken?	-	Every ___ minutes/hour? E.g. every 30 minutes for 1 hour.
☐ When should the experiment be conducted?	-	Hummingbirds feed best at dawn (5 am) and dusk (5pm).

3. Apparatus :

☐ What equipment/ lab apparatus is needed?	-	Sugar, water, kettle, measuring cups, funnels, bird feeders, cans of spray paint, hooks, a manual clicker counter, stop watch/ timer, notepad, etc...
☐ Is it available? OR Will it have to be made or improvised?	-	Most feeders are readily available in pet stores.
	-	Feeders can be made from plastic bottles.
☐ How must the apparatus be used?	-	Paint birdfeeders in appropriate colours, use the manual clicker counter to count the number of birds at each feeder.
☐ Use real specimens or models?	-	Real hummingbirds.
☐ Investigate in the field or in the lab? (based on resources – money, time, etc)	-	Hummingbirds are easier to count in the field, instead of capturing them to test them in the lab.

4. Treatment/ Trials

☐ Should there be repeats or trials? Always! (repeated measurements AND set-ups)	-	Count the number of birds at each feeder every day for at least 5 days.
☐ How many treatments to make?	-	Two red feeders, two yellow feeders and two green feeders = 3 treatments x 2 set-ups.
☐ Is there a control set-up?	-	The red feeders are the control set-up. OR a colourless feeder or one painted white.

5. Precautions

☐ Identify measures to be taken	-	Wait enough time after painting the feeders, so the paint is not toxic to the birds.
	-	Change the solution in the bird feeders everyday
☐ Are there any safety considerations?	-	Avoid harming or touching the birds?

Outline of the expected results and Interpretation (explanation) – based on example 2:

Expected results:

A) TABLE SHOWING THE NUMBER OF HUMMINGBIRD VISITS TO COLOURED FEEDERS AT 5:00 A.M.

Day	Number of hummingbirds visiting each feeder within 30 minutes.						
	Red 1	Red 2	Yellow 1	Yellow 2	Green 1	Green 2	Weather/ Brightness
1							
2							
3							
4							
5							
Total count							
Average							

B) Possible graph showing expected results

BAR GRAPH SHOWING EXPECTED AVERAGE NUMBER OF HUMMINGBIRD VISITS TO DIFFERENT COLOURED FEEDERS WITHIN 30 MINUTES ON MORNINGS.

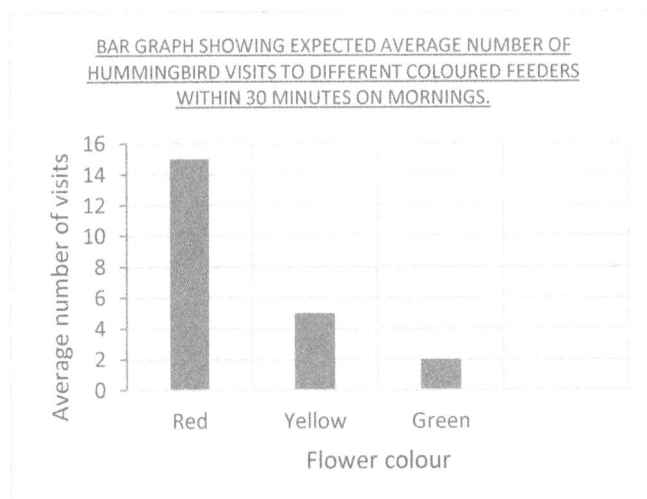

C) It is expected that hummingbirds will visit all of the feeders on any given morning. Furthermore, it is expected that a higher proportion of hummingbirds on average will visit the red feeders as shown in the expected graph.

If more hummingbirds visit the red feeders, then the hypothesis is accepted. However, if the results show that there is no preference in feeder colour, then the hypothesis is not accepted and some other factor is responsible for the original observations that "hummingbirds are often seen feeding on red coloured flowers in the garden".

Limitations:

In this experiment, by using feeders instead of naturally growing flowers, there is a chance that the paint colours - red, etc, may not be the same shade and hue as the natural flower colours. Additionally, the same bird may constantly be at one feeder, not giving others a chance to visit it. In the design of this experiment, it will be very difficult to watch all the feeders at the same time on the same day during the 5:00 to 6:00 morning period; so there may be the need for more observers or to introduce cameras and review the tapes.

LAB SKILLS IN CSEC BIOLOGY

Sample P&D Example 3

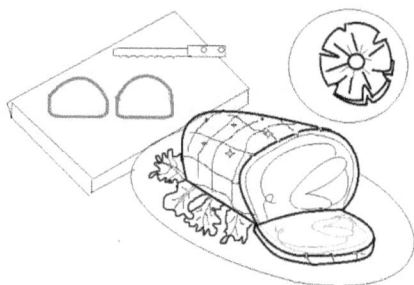

Observation:
Last Christmas while having a ham and hops party at home, Lucy observed her aunt putting pineapple slices on the ham sandwiches. Her aunt said that the pineapple had a natural chemical in it, which helps aid digestion of the ham by making it softer. Lucy decided to conduct a plan and design experiment using this information.

Developing the hypothesis: - *identify the manipulated and responding variables to create a statement.*
Manipulated variable –pineapple chemical/ enzymeResponding variable – texture of the meat.

Hypothesis: Natural chemicals (protease enzymes) in pineapple makes ham meat softer.

Aim: To investigate the effect of natural protease enzymes in pineapple on the texture of ham meat.

Apparatus and materials:

- 4 Beakers
- Measuring cylinder
- 5 Petri dishes with covers
- Scale balance
- 4 Syringes
- Bunsen burner
- Tripod stand
- Ruler
- Blender or juicer
- Labels
- Magnifying glass

- Forceps/ skewer
- Gloves
- Knife and cutting board
- Access to running water
- Ham meat (cooked)
- Fresh pineapple puree or juice
- Boiled pineapple juice
- A protease solution from the lab such as pepsin
- Distilled water

Method (*written in the active voice, present tense*)

1. Peel and cut a fresh pineapple. Using the juicer, obtain (at least 100ml) pineapple juice in a beaker labelled fresh pineapple juice.
2. Measure 50ml of the fresh juice, place in a beaker labelled boiled juice, and then boil it for 5 minutes.
3. Label 4 Petri dishes P1, P2, P3 and P4.
4. Using the scale, and clean cutting board, obtain 12 pieces of ham chunks each 10g in mass approximately the same sized cubes.
5. Using the gloves, place 3 pieces of ham in each of the 4 Petri dishes. Note the texture/ flakiness.
6. Add 20ml of the following solutions to each Petri dish:
   P1 – distilled water (note: this is the control)**P3** – boiled pineapple juice
   P2 – fresh pineapple juice**P4** – Pepsin solution.
7. After 30 minutes, observe the texture of each piece of ham; using clean gloves feel the pieces. Record the texture, firmness and other observations in an appropriate table.

Variables:

Manipulated variable – type of solution (presence of enzyme)
Responding variable – texture of the meat;
Constant variables –

- [ ] same volume of solution in each test tube;
- [ ] meat cubes left for 30 minutes each
- [ ] same mass of cubes in each treatment
- [ ] 3 cubes per Petri dish
- [ ] same depth Petri dishes

Expected Results:

TABLE SHOWING EXPECTED RESULTS OF A TEST TO INVESTIGATE THE PRESENCE OF ENZYMES IN PINEAPPLE WHICH MAKES MEAT SOFTER.

Test tube	Test solution with the ham	Expected Observations	
		Before 30 minutes	After 30 minutes
P1	Distilled water	Firm texture, meat springs back when touched.	Ham should be firm and spring back when touched.
P2	Fresh pineapple juice		Ham should be soft, pieces fall apart and flake off when touched.
P3	Boiled pineapple juice		Ham should be firm and spring back.
P4	Pepsin		Ham should be softer and pieces fall off when touched

Interpretation of Expected Results:

- Pineapple contains the enzyme bromelain, which is a natural mixture of two protease enzymes.
- Ham meat is made up of proteins, the substrates for protease enzymes.
- Bromelain is a meat tenderizer, effective in breaking down protein, such as collagen in steak. When this happens, the ham becomes softer.
- The ham with the fresh pineapple juice will be broken down, just as when using pepsin enzyme.
- Since pineapple juice has the bromelain enzyme, boiling it, would denature the enzymes, rendering it ineffective and causing the ham to remain firm.
- If the fresh pineapple juice makes the ham softer, then the hypothesis is accepted.

Limitations:

- When the enzymes act on the meat, there is the possibility that acidity increases. Acids may also increase the breakdown of proteins in meat.
- Ensure that the boiled pineapple is cooled properly so that temperature can remain constant throughout the experiment.
- The concentration of enzymes in pineapple juice may not be equivalent (it may be greater or lesser) to the pepsin enzyme, so the degree of softness may vary.
- Thirty (30) minutes may not be enough time for the enzymes to cause the meat to soften.

E. REFERENCES

TEXTBOOKS

Atwaroo-Ali, L. (2014). Biology for CSEC Examinations. Macmillan: Macmillan Publishers Limited.

Carrington, C., Agard, M. & Sealy, L. (1995). Biology: Skills for Excellence. Essex: Longman Caribbean.

Chinnery, L., Glasgow, J., Jones, M.,& Jones, G. (2001). CXC Biology. Cambridge: Cambridge University Press.

Morrison, K., Kirby, P. & Madhosingh, L. (2014) Biology for CSEC. Nelson Thornes

DOCUMENTS

CXC. (1997). Biology Resource Materials for Teachers: Book 1 – Orientation Module & Guidelines for School Based Assessment. Caribbean Examinations Council.

CXC. (2013). CSEC Biology Syllabus. Effective for Examination May-June 2015. Caribbean Examination Council.

WEBSITES

The impact of predation on life history evolution of Trinidadian guppies *Poecilia* reticulata
http://www2.hawaii.edu/~taylor/z652/ReznickEndler.pdf

Selection on Colour in Guppies. http://schoolbag.info/biology/living/113.html

Science at multiple levels http://undsci.berkeley.edu/article/howscienceworks_19

LAB GUIDES

Science Questions

Read

Experiment

Teamwork

Measure

Record Data

Analyser

Science Report

Submit Report

Feedback

LAB GUIDES

STUDENT NAME:_____ CLASS:_____

SCHOOL: _____ TEACHER:_____

TABLE OF CONTENTS			SKILLS				
Lab #	Page #	TITLE/AIM	ORR	D R	MM	A I	PD

_____

STUDENT NAME:_____ CLASS:_____

SCHOOL: _____ TEACHER:_____

Lab #	Page #	TITLE/AIM	ORR	D R	MM	A I	PD

TABLE OF CONTENTS — **SKILLS**

_____ CLASS:_____

NAME:_____**CLASS:**_____ **SKILL: MM/ORR**

TOPIC – SOILS/ EDAPHIC FACTORS   SYLLABUS OBJECTIVE: A 2.3

1.   INVESTIGATING THE WATER HOLDING CAPACITY OF DIFFERENT SOIL TYPES

AIM:_____

_____

Why do some areas flood and others do not? Which soil type is better for farming/ needs irrigation? This experiment investigates and compares the water holding capacity (a soil property) of two different soil types. It may be modified to include more soil types, if available.

APPARATUS and MATERIALS:

- Sand soil
- Clay soil
- Glass rod
- 2 Beakers
- 2 Filter papers

- 2 Funnels
- 2 Petri dishes
- Distilled water
- Spatula/ spoons
- 2 Measuring cylinders

- Electronic scale/ balance
- Clock
- Labels

DIAGRAM:

Fold filter paper in half

Fold into quarters

Open cone

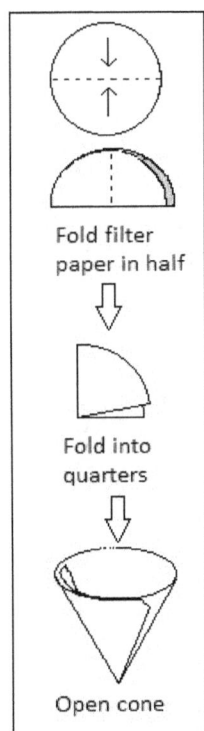

How to fold filter paper to fit into a funnel.

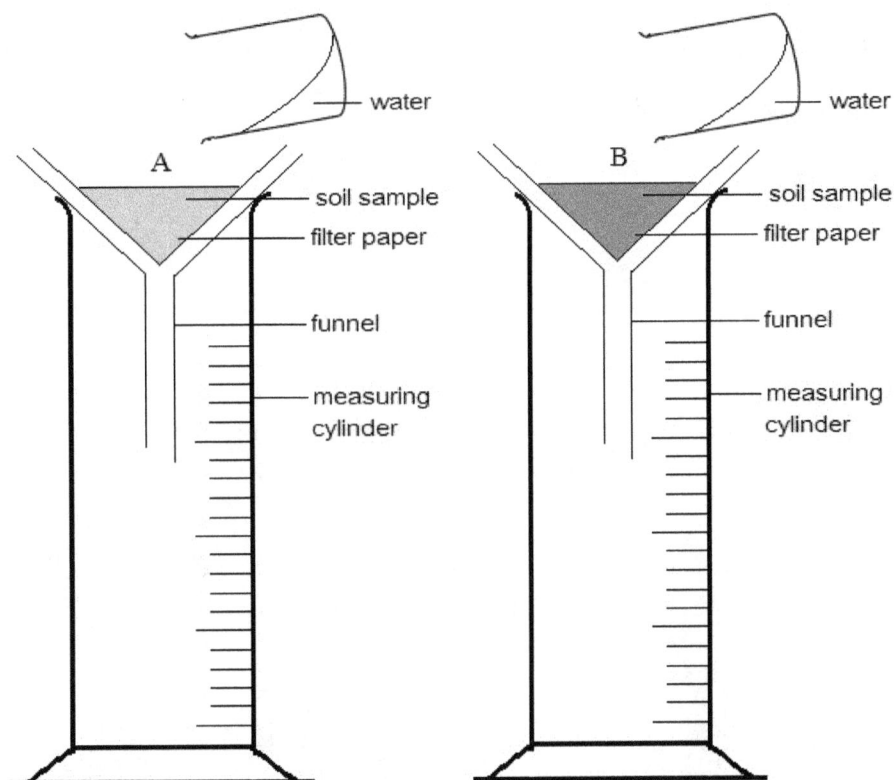

DRAWING SHOWING SET-UP OF APPARATUS TO INVESTIGATE THE WATER HOLDING CAPACITY OF SOILS.

INSTRUCTIONS:

1. Label 2 Petri dishes, Clay and Sand.
2. Using an electronic balance, place a Petri dish on the platter and zero the balance.
3. Measure out 20g of dry clay soil into the Petri dish Clay.
4. Repeat steps 1 and 2 for 20g of dry sand soil placing it in Petri dish Sand.
5. Using a measuring cylinder, measure $50cm^3$ of water and place each into beaker A and beaker B.
6. Prepare two sets of apparatus as in the diagram, with the filter paper in the funnel and the funnel in the measuring cylinder. Label each measuring cylinder sand and clay as appropriate.
7. Carefully place each soil sample into the correct filter paper.
8. Carefully pour the $50cm^3$ of water into each set of soil and allow it to drain for 10 minutes.
9. Observe and record the volume of water that drains through each soil type.
10. Note also the colour and texture of each soil type.
11. Calculate the volume of water retained by each soil sample.

Rewrite your method into past tense in the space below or on a separate page.

METHOD:

RESULTS: (*Add a title to the table and record your observations below.*)

TABLE SHOWING _____

	Soil Samples	
	Clay	**Sand**
Total volume of water poured (cm^3)		
Volume of water drained (cm^3)		
Volume of water retained by the soil (cm^3)		
Additional observations: - **Colour of soil** - **Texture of the soil**		

SAMPLE CALCULATIONS:

Volume of water retained in clay soil = Volume of water drained – Total volume of water poured

_____

_____

_____

_____

_____

_____

DISCUSSION

_____

_____

_____

_____

_____

_____

_____

_____

_____

_____

_____

_____

_____

_____

_____

_____

LIMITATIONS

_____

_____

PRECAUTIONS

_____

SOURCES OF ERROR

_____

CONCLUSION _(What was found out? Which soil type held more water and which soil drained better?)_

_____

_____

_____

REFLECTION _(How does this knowledge relate to your everyday life?)_

_____

_____

DISCUSSION QUESTIONS:

1. Identify the different soil types classified by scientists. What feature is used to classify them?

2. Based on the data recorded, which soil type retained more water?
3. Why does sand/clay retain more water? (Explain in terms of particle sizes and air spaces.)
4. Why does sand/ clay drain more water?

5. Why was the soil sample dry before adding the water?
6. What limitations and/ or sources of error was encountered in collecting results for this experiment?

NOTES: SOILS

Soil – is a mixture of the remains of organic material (remains of plants and animals) and inorganic material like minerals weathered from large rocks. Soil is very important for life since it allows plants to anchor and obtain water and nutrients. It is a habitat for animals such as termites and earthworms. Soils also play a role in purification of water by filtering out wastes and pollutants.

Soil types can be classified based on the size of the particles in it. Clay type soils have the smallest particles of diameter 0.002mm, silt particles are 0.002-0.02mm and sand particles are 0.02mm-2.0mm.

In general soils from different parts of your country are mixtures of different particles, textures and compositions based on the types of bedrock from which they are formed. Ultimately the soil properties such as water holding capacity, soil drainage, nutrient content, fertility and possible uses are affected by the soil type.

*   **Sandy soils** – consist of large particles. The spaces between the particles trap a lot of air but the soil is very porous, so water drains quickly. Sandy soils are usually nutrient poor.
*   **Clay soils** – have tiny particles, because of this water holds easily by capillarity and fill up the pores making drainage difficult. Clay soils tend to swell and stick together. They are also reasonably fertile.
*   **Loam soils** – are a mixture of sand, clay, silt and humus (remains of organic materials). They are very fertile and excellent for plant growth.

SUGGESTED MARKSCHEME – MANIPULATION AND MEASUREMENT (MM)

Criteria (MM)	Marks	
Using the measuring cylinder (1 mark each)		
☐ Placement on a flat surface/ table top		
☐ Readings taken at eye level		
☐ Readings taken at the bottom of the meniscus	5	
☐ All liquid poured out completely from cylinder		
☐ Cylinder is cleaned and washed between uses and at the end		
Filter paper - Correct folding of the filter paper and placement in the funnel	2	
Using the scale balance (1 mark each)		
☐ Clean the platter of the scale before weighing		
☐ Zero the scale before adding Petri dish and before adding the soil sample	3	
☐ Obtaining the correct mass of sand or clay		
TOTAL	10	

ADDITIONAL INVESTIGATIONS WITH SOILS

(A) <u>INVESTIGATING THE COMPONENTS OF SOIL – SOIL PROFILE/ SEDIMENTATION TEST</u>

<u>Instructions:</u>

1. Place a sample of garden soil into a large measuring cylinder or gas jar (approximately ¼ full).
2. Add tap water until within 5cm of the top of the apparatus.
3. Using a glass rod, stir the mixture thoroughly. Then leave to stand until the particles settle.
4. Observe and record the layers present. (Use a ruler to estimate the percentages of each layer.)
5. Obtain other soil samples and compare them.

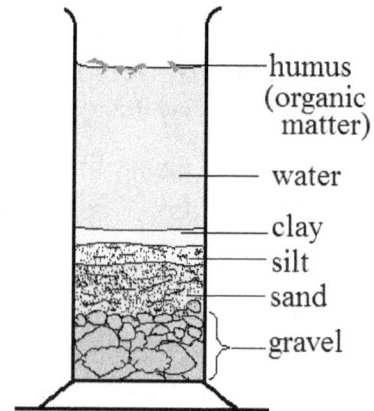

— humus (organic matter)

— water

— clay
— silt
— sand
— gravel

A TYPICAL SOIL PROFILE

Note: Larger, denser particles sink to the bottom layers (quickly), clay may remain suspended in the water, but given enough time (say 24 hours) will settle. Organic materials (humus) is usually at the top.

(B) <u>INVESTIGATING THE AMOUNT OF AIR IN A SOIL SAMPLE</u>

    a.   <u>Mass method - Instructions:</u>

1. Fill a beaker with soil, then weigh the beaker and soil. Record initial mass.
2. Slowly add water, until the water is level with the top of the soil.
3. Re-weigh the beaker with soil and water. Record final mass value.
4. Calculate the percentage of air in the soil sample as:
$$\frac{Final\ mass\ -Initial\ mass}{Initial\ mass} \times 100$$
5. Repeat steps 1 – 4 for two more samples of the same soil to obtain the average percentage of air.

    b.   <u>Volume method - Instructions</u>

1. Place 50 cm³ of soil in a measuring cylinder.
2. Add 50 cm³ water to the measuring cylinder of soil.
3. Stir the mixture gently until no more bubbles escape.
4. Record the volume of the mixture as Mcm³.
5. Calculate the volume of soil air as: 100 cm³ – Mcm³.

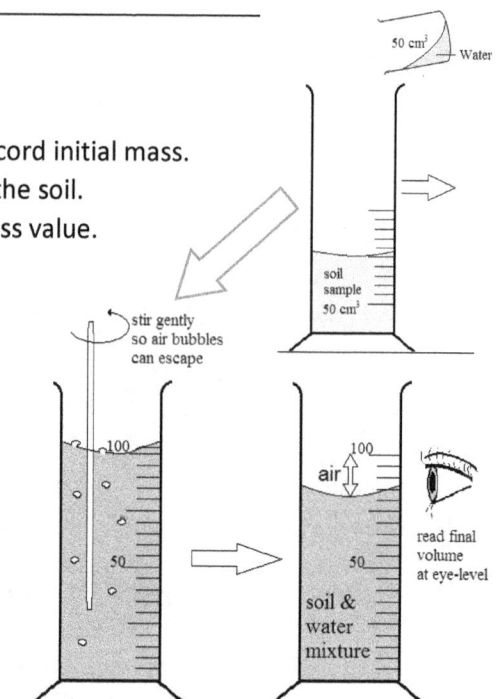

(C) INVESTIGATING THE WATER CONTENT OF SOILS – DRYING METHOD

<u>Instructions:</u>

1. Weigh a sample of soil using a scale.
2. Heat the sample for 10 minutes at 90°C in an oven. Cool and re-weigh the sample.
3. Reheat the soil again for several minutes, and re-weigh again.
4. Repeat step 3 until there is no further loss in weight.
5. Calculate the percentage of water in the soil sample using the formula:
$$\frac{Mass\ of\ wet\ soil\ -final\ mass\ of\ dry\ soil}{mass\ of\ wet\ soil} \times 100$$

NAME:_____ **CLASS:**_____ **SKILL: ORR**

TOPIC: ECOLOGY   SYLLABUS OBJECTIVES: A2.1, 3.1, 3.5

2.   AN ECOLOGY STUDY OF A POND AND ITS IMMEDIATE SURROUNDINGS AT THE WILD FOWL
      TRUST, POINTE-A-PIERRE

AIM: 1. TO INVESTIGATE A POND ECOSYSTEM AND ITS IMMEDIATE SURROUNDINGS AT THE POINTE-
      A-PIERRE WILD FOWL TRUST.

      2. TO RELATE THE BIOTIC AND ABIOTIC FACTORS THAT AFFECT A POND ECOSYSTEM AND THE
      SURROUNDING HILLSIDE.

3._____

As a young scientist visiting an area, do you wonder what organisms live there? Where are they
found? Or how many are there? The purpose of this lab is to introduce you to some of the
sampling methods and apparatus used to investigate a freshwater pond and a section of hillside
or other forested area. In observing, identifying and recording the biotic and abiotic
components a better understanding of ecological interactions can be gained.

APPARATUS & MATERIALS:

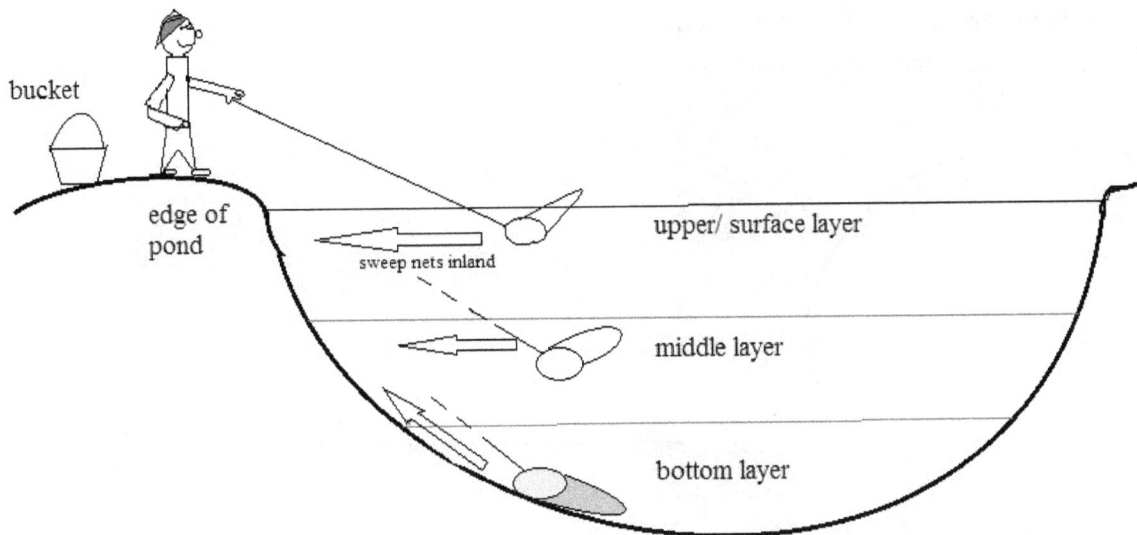

bucket

edge of pond

sweep nets inland

upper/ surface layer

middle layer

bottom layer

DIAGRAM 1 SHOWING THE POND DIP METHOD FOR SAMPLING THE POND.

METHOD:

(A) Pond dip sampling:

1. Carefully examine the surface of the water for animals that cannot be caught in the net. Note the abiotic factors such as, temperature of water, turbidity and pH.
2. Using an appropriate mesh aquatic net, reach out as far as possible (without falling over) and place the net just under the surface of the water.
3. Sweep the net towards the edge of the pond, lift it out and empty the contents into the clean water in the bucket. Repeat 2 more times.
4. Use the pond identification cards to identify and record each organism collected.
5. Repeat Steps 1 to 4 for the middle and bottom layers of the pond, using the appropriate buckets/nets.
6. Construct four food chains and use them to draw a food web to show the feeding relationships in the pond.

(B) Line transect (systematic sampling)

7. Run a 50m measuring tape (line transect) from the top of a slope to the bottom where the pond is located.
8. Place a flag at 10 m intervals, then a 0.5 m² quadrat at each interval in alternate positions.
9. Carefully observe and record the estimated percentage of different plant species in each quadrat.

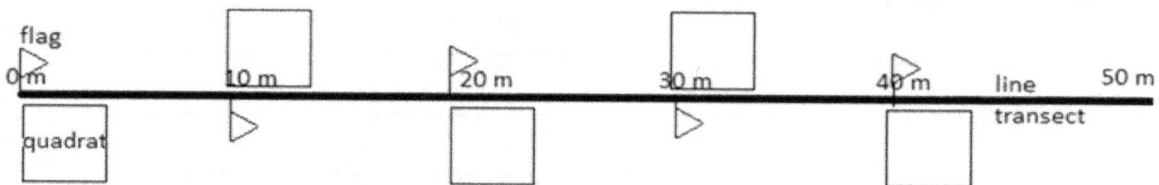

DIAGRAM 2 SHOWING THE LINE TRANSECT SAMPLING AREA/ SYSTEMATIC METHOD

(C) Quadrat sampling (random sampling)

10. Randomly throw a quadrat on the study area and note the number of each species of plant seen.
11. Repeat step 10, three more times.

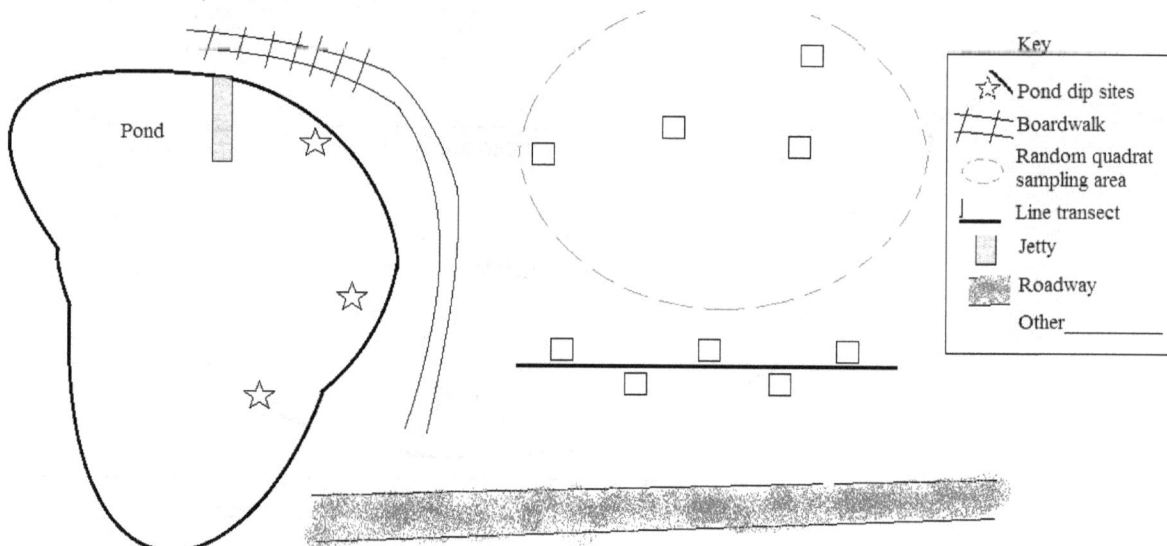

DIAGRAM 3 SHOWING SKETCH OF THE ENTIRE STUDY SITE AND SAMPLING AREAS

RESULTS:

(A) RESULTS OF THE POND DIP SAMPLING

TABLE 1: SHOWING ORGANISMS AT DIFFERENT WATER DEPTHS

Depth of pond	Names of organisms observed (and food source)
Surface	
Middle	
Bottom	

TABLE 2: SHOWING SOME ABIOTIC FACTORS OF THE POND

Abiotic factor	Information recorded
Turbidity	(clear/ cloudy)
pH of water	(acidic/ alkaline)
Temperature (°C)	

Draw FOUR aquatic food chains based on the feeding relationships in the pond; see example below.

Detritus ⟶ leech ⟶ fish ⟶ anhinga bird
(Producer)   (primary consumer)   (secondary consumer)   (tertiary consumer/top carnivore)

1. _____

2. _____

3. _____

4. _____

DRAWING SHOWING AN AQUATIC FOOD WEB OF ORGANISMS IN THE POND

(B)   RESULTS OF THE LINE TRANSECT SAMPLING – SYSTEMATIC SAMPLING

TABLE 3: SHOWING ORGANISM PRESENT ALONG LINE TRANSECT (SYSTEMATIC SAMPLING)

QUADRAT at DISTANCE (m)	SPECIES OF ORGANISMS PRESENT (%)					Total Plant cover %	Total Bare ground %	Sun-light %	Soil type, texture and notes.
	Plant A	Plant B	Plant C	Plant D	Other				
0									
10									
20									
30									
40									
50									

Sketch of plant leaf outlines + Description of texture, size, colour				

*Note: Total plant cover = % Plant A + % Plant B + % Plant C + % Plant D + % Other

Total bare ground = 100% - Total plant cover %

Sunlight % - indicate how much sunlight is coming in overhead. 0% means you are under a shady tree.

(C) RESULTS OF THE RANDOM QUADRAT SAMPLING

TABLE 4: SHOWING THE PLANTS FOUND IN EACH QUADRAT DUE TO **RAMDOM SAMPLING**.

Plant Species	Number of organisms in Quadrat				Average no. of organisms	Population Density (plant species/m^2)	Frequency (%)
	1	2	3	4			
A							
B							
C							
D							
Example: Sweethearts	0	5	7	4	(0+5+7+4)/4 = **4**	8	75 *(present in three out of four quadrats)*

CALCULATIONS:

1. **Sample calculation for population density**

For species "Sweethearts":

Size of quadrat used (m^2): __0.5 m^2_____

Average number of sweetheart plants in area of 1 quadrat (0.5 m^2)= ___4___

So the number of sweetheart plants in a unit area of 1.0 m^2 = _8___

Therefore the population density of sweet heart plants is = _8 sweetheart plants / m^2_

2. **Sample calculation for species frequency**

Frequency of Sweethearts (%)= # of quadrats with Sweethearts present/ total # of quadrats

= 3 quadrats/ 4 quadrats total

= 75%

ADDITIONAL OBSERVATIONS

Weather: _____

Time of sampling: _____

Interesting birds seen: _____

Bird activities: _____

Reptiles/ Amphibians: _____

Reptilian/Amphibian
activities: _____

Insects encountered: _____

Plants identified: _____

Condition of pathway: _____

Odours in the air: _____

Sounds: _____

DISCUSSION_____

_____

_____

_____

_____

_____

_____

_____

_____

_____

_____

_____

_____

_____

_____

_____

_____

_____

_____

_____

_____

_____

_____

_____

_____

_____

_____

_____

_____

CONCLUSION *(Identify the ecosystem and habitats investigated and state what was found out in terms of abiotic factors affecting biotic factors. State the apparatus used for the study.)*

_____

_____

_____

_____

_____

_____

_____

REFLECTION *(Explain how knowledge gained in Ecology relates to YOUR everyday life.)*

_____

_____

_____

_____

_____

_____

_____

_____

DISCUSSION

1. Define the term "ecology".
2. Identify the characteristic features (major abiotic and biotic factors) of the 2 habitats studied in this lab - pond and surrounding terrestrial area.

3. What did the food web of the pond show? Identify the names of the producers, top carnivore and its food sources.
4. Why is it important to have more than one food source?
5. With respect to the pond dip method
   a. Why was a net used?
   b. Why were the 3 layers of the pond sampled?

6. With respect to sampling with quadrats
   a. Briefly compare how they were used in systematic and random sampling.
   b. Why was the quadrat thrown several times for the random sampling?
   c. What kinds of organisms are most suitable to sample using a quadrat?

7. Based on the results collected along the transect, answer the following:
   a. What resources do plants compete for? Explain why.
   b. Is there a relationship between the type of plants species and the amount of sunlight available?
   c. Using values for the population density, explain which plant favours full sunlight and which is better adapted to shade.
   d. What was observed about the type of plant species and location on the hillside/ slope? Explain which soil conditions (if any) affected plant distribution.

CONCLUSION (Write a statement about **what was found out during the day's activities** with respect to sampling an ecosystem AND the relationship between biotic and abiotic factors in an environment.)

SAMPLE MARK SCHEME - ORR CRITERIA				Max. Mark	Teacher Mark
O – Observations	s o c/d	• Significant changes noted • Original and final conditions compared • Control noted OR diagram		3	
RTG – Recording	t u e	**TABLE** • Title – above, in capitals, underlined (1) • Column & row headings(with units) (1) • Enclosed and neat (1)	t u p · **GRAPH** • Title below, capitals, underlined (1) • Both axes labelled with units (1) • Accurate plots (1)	3	
R – Reporting	a r g s	• Aim in capital letters (1) • Reflection appropriate (1) • Acceptable language and expression – subject-verb agreement/ grammar (1) • Spelling correct throughout with 0-3 errors (1) *Note: > 3 grammatical errors = 0 marks. AND > 3 spelling errors = 0 marks*		4	
TOTAL				10	

ECOLOGY AND SAMPLING ECOSYSTEMS NOTES:

Ecology is the scientific study of organisms and their environment. Ecological studies can identify the **distribution (where)** and **abundance (how many)** of organisms, including **relationships** between biotic components or biotic-abiotic **interactions**. When an ecological study is done, it is impossible to count all plants and animals present or look at every single area, so a **sample** is conducted. Samples are "pieces or portions" that are representative of the whole study area.

An ecosystem can be as large as a biosphere or as small as a dot (a bacterial colony). Ecosystems refer to many organisms and how they relate to each other and their environment. Some **types of ecosystems** found on the planet are:

HABITAT – describes the place an organism lives out its life. Habitats are described specific to one organism.

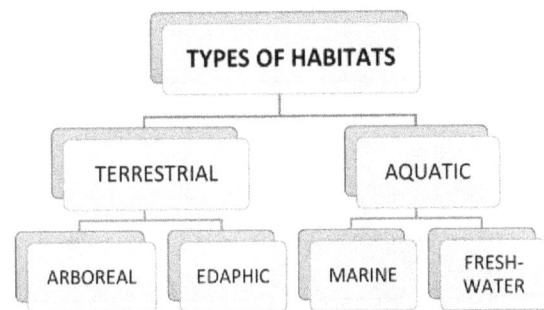

TYPES OF HABITATS

TERRESTRIAL   AQUATIC

ARBOREAL   EDAPHIC   MARINE   FRESH-WATER

1. Land/ Terrestrial ecosystems – forests, savannahs or deserts.
2. Freshwater ecosystems – ponds, rivers.
3. Wetland ecosystems – swamps, marshes, estuaries.
4. Marine ecosystems – coral reefs, oceans.

Each ecosystem contains different organisms; which are **adapted to survive the abiotic factors** in their habitats. Some examples are:

- Water Lilies have special air tissues to take oxygen down to the roots **under the water in anoxic (waterlogged and toxic) soils**.
- The Greater Water-boatman (a swimming insect) has "oars" made of long hind legs and dense hairs to propel it through **the water**.
- Vines in forests are adapted to climb up into the canopy **to get sunlight and avoid being shaded**.

Even though abiotic factors may affect organisms; organisms themselves can alter their environment and that of other species. For example, aquatic algae and plants increase oxygen levels in ponds during the day due to photosynthesis, but lower the oxygen levels during the night as respiration increases. The lower limit of oxygen determines how many aquatic organisms survive there.

Organisms are classed as producer, consumer, predator, prey, host or parasite based on their feeding relationships. Biotic interactions are represented by feeding relationships shown in food chains or food webs. They both show the direction of energy flow from producers to consumers.
  a. **Food chains are linear** feeding relationships showing organisms and only one food source.
  b. **Food webs** show the interactions between many organisms; **linking all possible food sources** of consumers at each trophic level. It is essentially many food chains joined together.

SAMPLING TECHNIQUES AND APPARATUS

To understand an ecosystem, the community structure must be determined. This entails identifying the habitats and populations therein and determining their abundance. All the places in an environment where an organism is found living is called its **distribution**. Based on the size of the area, habitats within and the time available, different types of sampling may be done. These factors also determine which sampling techniques to use.

Types of sampling

1. Systematic sampling	2. Stratified sampling	3. Random sampling
• The area is not very large. • The area is varied along a gradient (down a hill or across a river). • Samples are taken at fixed intervals along a transect line.	• There are different smaller areas within a larger habitat. • For example, a pond and a forested area within a larger forest. • Each area can be studied using different sampling techniques.	• The area is fairly uniform. • The area is very large. • There is limited time to do the study. • Every part of the sample area has an equal chance of being sampled.
Side view of a hill slope.	Aerial view of forest with a pond.	Aerial view of a uniform forest.

Ecologists use various **sampling techniques/ apparatus** to study abundance and distribution of organisms. For small areas this can be recorded with a simple sketch map and lists, but for larger areas the following may be used:

(A) Sampling plants/ terrestrial organisms that do not move (sessile or sedentary organisms):

1. **Transect**

a. **Line transect** - A line or rope, even a tape measure run along the ground/ surface between two points (sticks or trees).

   i. All the plant species that touch the line are recorded (continuous sampling), or,

   ii. at regular intervals, say 10m, use quadrats and record plants. This is systematic sampling.

b. **Belt transect** – two line transects about 1m apart.

   i. Record organisms in that width or area continuously, or at intervals.

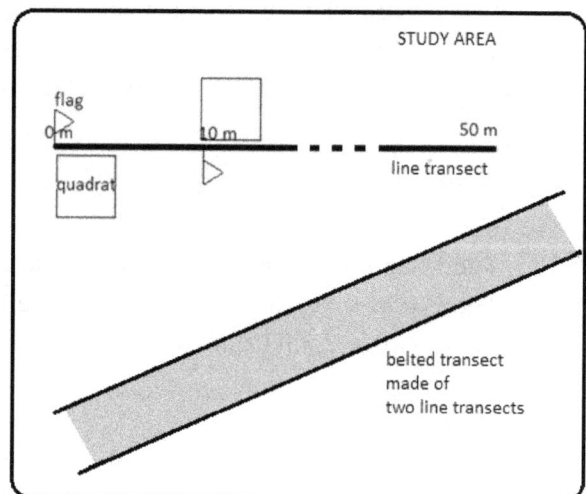

2. **Quadrat** – A square frame (wooden, plastic or metal) of known size e.g. 0.25m² or 1m². The internal frame can be further divided with string into smaller grids or segments.

   i. Quadrats can be placed systematically every 10m along the line transect. All organisms in the quadrat can be count and recorded and abundance found.

   ii. Quadrats can be thrown randomly over the shoulder multiple times when studying a uniform area.

   iii. Random quadrat sampling can be done by using a table of random numbers, given a map of the area and then going to those coordinates. At least 25% of the total area should be sampled.

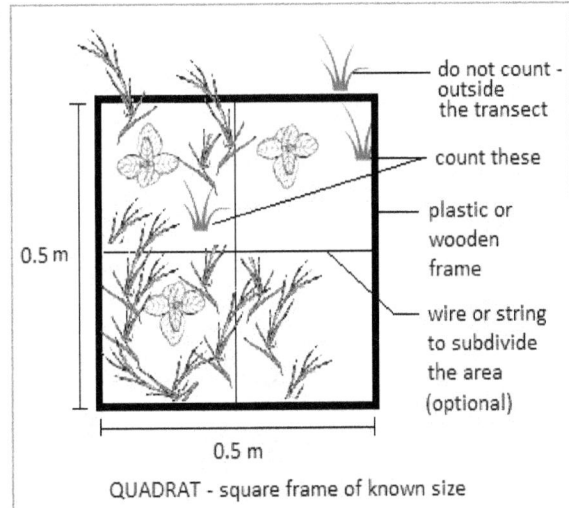

QUADRAT - square frame of known size

(B) **Sampling animals.** (Animals are usually **mobile or secretive** and can hide.):

   1. **Traps** – depending on the size and type of animals trying to be caught:

a. **Pooter** – suction device to trap small invertebrate/ mobile organisms in a jar	b. **Pitfall trap** – for crawling insects that live close to or on the soil surface
	Note: Leave overnight, for 24hrs; but remove traps after!
c. Simple **trap boxes with food**/ attractive substances (dry bedding) for organisms of varying sizes Note: Ensure animal is removed unharmed from trap within 24 hours.	**Tullgren funnel** – for soil organisms trying to get away from heat and light

2. **Nets** – depending on the habitat studied, different type (materials and size) nets can be used to catch organisms for closer examination.

a. **Sweep net** (canvas) – for organisms in long grasses/ **butterfly net** (mesh) – for flying organisms.	b. **A-frame net/ fishing net** – for aquatic organisms.	c. **Plankton net** – for collecting plankton only, discarding most of the water.
180° arc canvas or mesh rotate handle to close opening Note: canvas is stronger to sweep through grass/ undergrowth	long handle strong mesh cloth	towing line from boat key ring string wire nylon or mesh net plastic bottle

3. **Beating trees/branches with sticks** then collecting and counting the organisms that fell out.
4. **Digging in the mud/ sand** for earthworms, crabs, chip-chip and other invertebrates.
5. **Collecting** night flying organisms **with a white sheet and a light bulb**.
6. **Mark-release-recapture:** capture a sample of organisms (10 deer), mark them with a non-toxic substance or tag/tracking device, release them and let them mix with the original population. At a later date, sample again; if 50% of the new sample is marked (5 out of 10 deer), then it can be expected that the original sample was 50% of the whole population. So the total deer population was the original ten deer multiplied by two, that is, 20 individuals.

NOTE: ALWAYS <u>handle organisms carefully</u> and <u>return them safely</u> to their habitat after recording.

<u>**ECOLOGICAL CALCULATIONS:**</u>

1. **Abundance** – how many species are found in an ecosystem. Species abundance can indicate the contribution of each species to the community. The following are all measures of abundance:
   a. **Count** – simply counting the number of organisms in your study area can indicate population size, e.g., all of the poui trees on the school compound.
   b. **Density** = number of plants in a given area (total area of quadrats) e.g., all the plants in the area occupied by five quadrats (of 0.25 m²) = 200 plants/1.25m².
   c. **Relative density** = the density of a species of plant as a percentage of total density e.g. the relative density of fine leaf grass is [(50 grasses/1.25m²)/(200 plants/1.25m²)] x 100 = 25%.
   d. **Species frequency** - probability of finding a given species within any one throw of a quadrat. For example, if a fine leaf grass plant is found in only four out of ten quadrats sampled, its frequency is 40%.
   e. **Species cover** – proportion of ground covered by any one species given the area of the quadrat. This value can be multiplied to represent proportion of entire study area.
2. **Distribution** – Patterns of species distribution can be described as either (1) random, (2) regular/uniform, or (3) clumped (due to favourable environments or dispersal mechanisms).

NAME:_____   **CLASS:**_____   **SKILL: MM/ORR**

TOPIC: CELLS/ OSMOSIS   SYLLABUS OBJECTIVE: B 1.6, 1.7

3.   INVESTIGATING OSMOSIS USING CUCUMBER AND RAISINS

AIM:_____

_____

Have you ever made chow and wondered why water collected in the bowl after some time and the fruits got soft? After coming home from the hot market, how can you rejuvenate the lettuce you just bought? This experiment introduces you to the process of osmosis in plant cells. Cucumbers are exposed to salt crystals (a hypertonic situation) whereas raisins are exposed to distilled water (hypotonic solution).

APPARATUS and MATERIALS:

- Fresh cucumber
- Raisins (about 10)
- Distilled water
- 2 Petri dishes
- 2 Small beakers
- Knife/scalpel
- Spatula

- White tile
- Measuring cylinder
- Salt
- Paper towel
- Labels
- Stop clock/ watch

DIAGRAM:

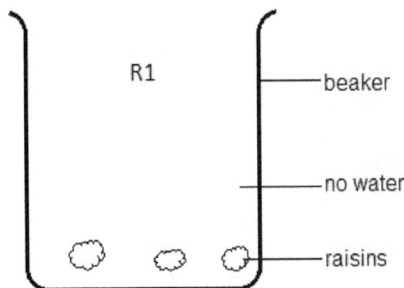

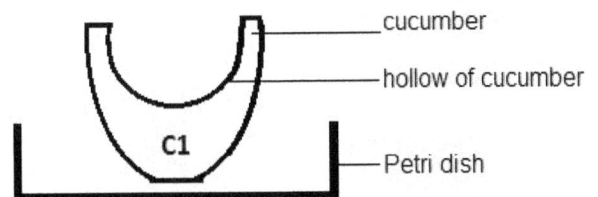

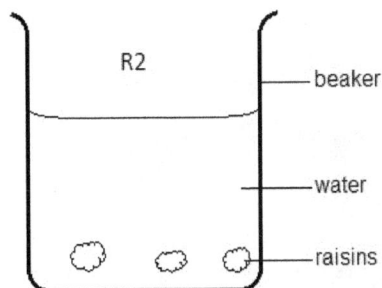

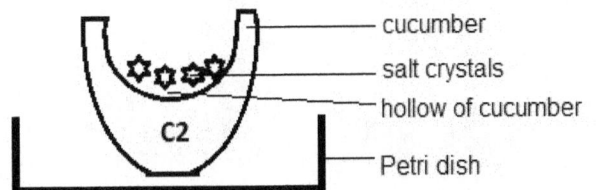

DRAWING SHOWING RAISINS IN WATER

DRAWING SHOWING SET-UP OF CUCUMBERS

INSTRUCTIONS:

1. Label 2 small beakers R1 and R2.
2. Place 3 raisins in each beaker R1 and R2, respectively.
3. Measure and add 40ml of water to the raisins in R2.
4. Observe and record the appearance of the raisins in each beaker at the start and after 25 minutes.
5. Cut a cucumber in half, cut the bases of each half horizontally so that each piece can stand upright.
6. Carefully use the spatula to scoop out the seeds but do not go through to the base.
7. Use a tissue and dry the centre of each half.
8. Label one half C1 and the other half C2.
9. Add 2 heaping spatulas of salt to the centre of C2.
10. Observe and record the appearance of the hollows of each cucumber at the start and after 25 minutes.

Rewrite your method in the past tense in the space below or on a separate page.

METHOD:

RESULTS: *(Add a title to the table and record the observations below.)*

TABLE SHOWING _____

Experiment	Observations (description of colours, sizes and textures)	
	At the start	After 25 minutes
R1 – Raisin (control)		
R2 – Raisin (in water)		
Cucumber C1 (without salt/ control)		
Cucumber C2 (with salt)		

Include a LABELLED drawing to **compare R1 and R2 at the end of the experiment**.
ANNOTATE to show the direction(s) of water movement into or out of tissues, if any.

TITLE: _____	TITLE: _____

DISCUSSION_____

_____
_____
_____
_____
_____
_____
_____
_____
_____
_____
_____
_____
_____
_____
_____
_____
_____

LIMITATIONS_____

_____

PRECAUTIONS_____

_____

SOURCES OF ERROR_____

_____

CONCLUSION *(What happened when plant cells are (i) placed in pure water and (ii) exposed to salt?)*

_____
_____
_____

REFLECTION *(Explain how this knowledge relates to your everyday life.)*

_____
_____
_____

DISCUSSION QUESTIONS

1. Define osmosis.
2. State what happens to plant cells placed in pure water (high water concentration – hypotonic).
3. State what happens when placed in a hypertonic solution (low water concentration) e.g. salt.
4. Will plant cells burst when placed in pure water? Why or why not?
5. Identify the high and low water concentration areas in the raisin set-up to explain the results.
6. Explain the observations for the cucumber set-up in terms of direction of water movement.
7. Identify the controls in each set-up. Why were they used?
8. Explain any precautions taken while setting up the cucumbers.
9. Is there any way to improve the results? (Hint: measuring change in mass over time.)

NOTES - OSMOSIS

- Osmosis is a special example of diffusion. It is the passive movement of **water molecules** through a **selectively permeable membrane** from a more dilute solution to a more concentrated solution.
- **Cell membranes** are described as selectively permeable since they allow the passage of water and also certain solutes (dissolved substances) but not other large molecules.
- The inside of the raisin cells had a low water concentration so water moved into those tissues.
- The cucumber cells had a higher water concentration than the salt crystals, so water left the cucumber cells, going from high to low water concentration by osmosis.
- The <u>control</u> for both experiments was used to ensure a comparison and that at the end of the 25 minutes the changes were due to the concentration of the surrounding water in the raisins and the salt in the cucumber hollow and not some other factor of the tissues themselves.

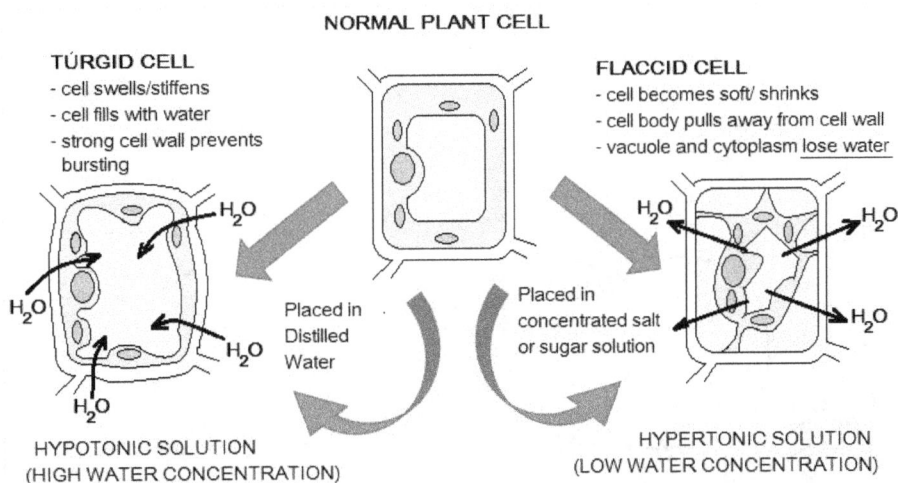

NORMAL PLANT CELL

TURGID CELL
- cell swells/stiffens
- cell fills with water
- strong cell wall prevents bursting

FLACCID CELL
- cell becomes soft/ shrinks
- cell body pulls away from cell wall
- vacuole and cytoplasm lose water

H_2O

Placed in Distilled Water

Placed in concentrated salt or sugar solution

H_2O

HYPOTONIC SOLUTION
(HIGH WATER CONCENTRATION)

HYPERTONIC SOLUTION
(LOW WATER CONCENTRATION)

SAMPLE MARK SCHEME - ORR GENERAL CRITERIA			Mark	T. Mk.
O – Observations	s o c/d	• Significant changes noted – change in texture/ size/ moisture • Original and final conditions compared – for all situations • Control noted OR diagram – comparisons of the raisins at the end	3	
RTG – Recording	t u e	**TABLE** • Title above, in capitals, underlined (1) • Column & row headings(with units) (1) • Enclosed and neat (1)	t u p	**GRAPH** • Title below, capitals, underlined (1) • Both axes labelled with units (1) • Accurate plots (1)	3	
R – Reporting	a r g s	• Aim in capital letters (1) • Reflection appropriate (1) • Acceptable language and expression – subject-verb agreement/ grammar (1) • Spelling correct throughout with 0-3 errors (1) *Note: > 3 grammatical errors = 0 marks. AND > 3 spelling errors = 0 marks*	4	
TOTAL			10	

NAME:_____ CLASS:_____ SKILL: **MM/ORR**

TOPIC: CELLS/ OSMOSIS   SYLLABUS OBJECTIVE: B 1.6, 1.7

4.   INVESTIGATING OSMOSIS USING POTATO STRIPS IMMERSED IN DIFFERENT SOLUTIONS

AIM:_____

_____

Which do you prefer, soft potato chips or firmer, crisp potato chips? Did you know that soaking potato chips in water with or without salt can change the texture of the chips? Potato is made up of plant cells that can exhibit the effects of osmosis when immersed in different solutions. In this experiment the effect of different solutions on the length (and mass) of potato strips will be investigated.

APPARATUS and MATERIALS:
- Potato (remove peel)
- Cork borer* (opt.)
- 2 Petri dishes
- Distilled water
- Salt water solution
- Ruler
- Forceps
- Knife
- White tile
- Measuring cylinder
- Stop clock/ Watch
- Paper towel
- Labels

DIAGRAM:

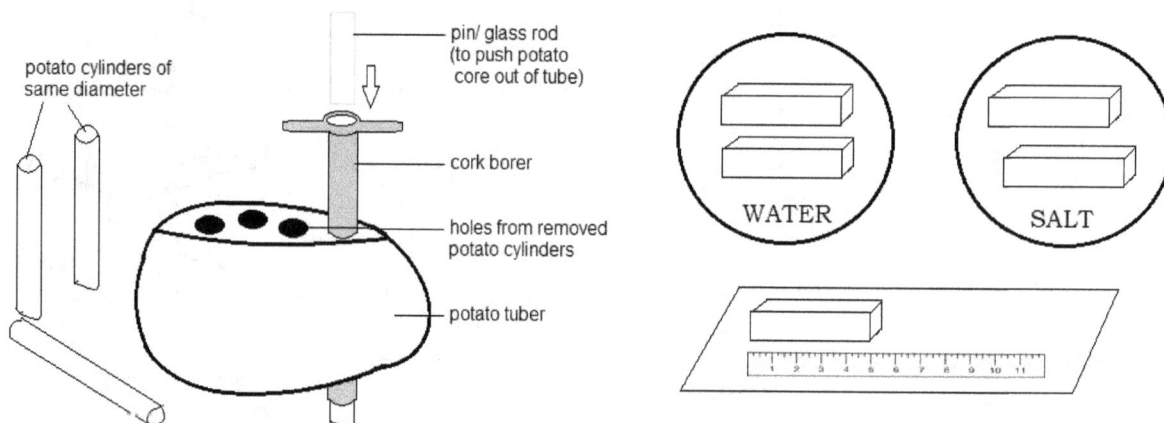

DIAGRAM SHOWING THE SET UP OF APPARATUS TO INVESTIGATE OSMOSIS.

INSTRUCTIONS:
1. Cut four potato strips of the same size (1cm x 1cm x 5cm), or use a cork borer to obtain four potato strips of same diameter and trim to 5 cm. Dry them on a paper towel.
2. Label one Petri dish 'water' and the other one 'salt solution', and add 30ml of each liquid.
3. Note the initial textures of each strip then, then add two to each dish. They should be completely immersed under the solutions.
4. Every 5 minutes, take out the strips, dry them and measure the length to the nearest mm.
5. Record the lengths in the table, along with the final texture of the strips. Find the averages.
6. Draw a line graph of the change in average lengths of the potato strips over time.

ALTERNATIVE MEASUREMENT – For each solution, record the average initial mass of the strips and the average final mass of the strips after 1 hour of complete immersion in the solutions.

RESULTS: *(Add a title to the table and collect the measurements, then plot a graph.)*

TABLE SHOWING _____

Time (mins)	Length of strip in water (cm)			Texture of strip in water	Length of strip in Salt solution (cm)			Texture of strip in salt solution
	1	2	Avg.		1	2	Avg.	
0								
5								
10								
15								
20								

Graph - *Plot on a separate graph page or below.*

Remember to put a TITLE and scale, label the axes with units of time (mins) or size (mm).

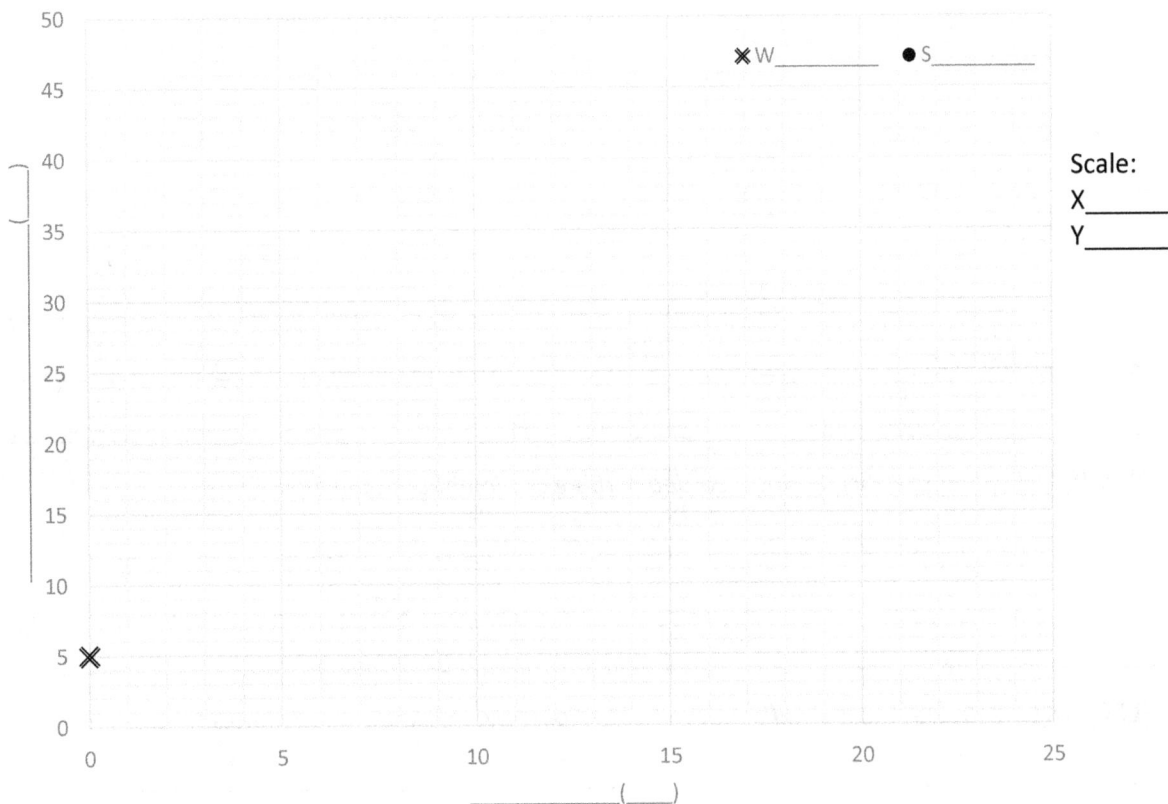

TITLE: _____

DISCUSSION_____

_____

_____

_____

_____

_____

_____

_____

_____

_____

_____

_____

_____

_____

_____

_____

LIMITATIONS_____

_____

_____

PRECAUTIONS_____

_____

SOURCES OF ERROR_____

_____

CONCLUSION (*What specifically was recorded in this experiment about osmosis?*)

_____

_____

_____

REFLECTION (*What was most significant to you? Explain how this knowledge relates to your life.*)

_____

_____

_____

NAME:_____CLASS:_____ SKILL: DR____

TOPIC: PHOTOSYNTHESIS   SYLLABUS OBJECTIVE: B 2.3

5.   INVESTIGATING THE EXTERNAL FEATURES OF A HIBISCUS LEAF

AIM:_____

_____

Leaves are the main organs of photosynthesis, they are made up of various parts all working together to help make food for the plant. This experiment allows you to observe a hibiscus leaf carefully and record important features. The skill of drawing specimens will also be developed.

APPARATUS AND MATERIALS:
- Leaves - hibiscus
- Magnifying glass
- White tile
- Labelling and annotation guide
- Mechanical pencil/ sharp pointed pencil

- Sharpener (Optional)
- Clean eraser
- Ruler
- White drawing paper

INSTRUCTIONS:
1. Obtain a leaf from a Hibiscus plant.
2. Examine its external features such as the shape, size, colour, texture/ thickness, location and arrangement of the veins, edges and petiole.
3. Make a large, labelled drawing of the leaf.

   (LABELS – leaf apex/ tip; edge/ margin; lamina/leaf blade; main vein/midrib; lateral veins/ net veins/ branching veins; petiole/ leaf stalk)

4. Using the guide (Table 2), add **at least 3** annotations (description and functions of labelled parts and relate to photosynthesis).
5. Calculate the magnification of your drawing and state it in the title of the drawing. Use the formula:

   Magnification of drawing = size of drawing/ real size of leaf

METHOD: *(Rewrite your method in past tense in the space below.)*

_____

_____

_____

_____

_____

RESULTS: (*Make a large, labelled, annotated drawing of the Hibiscus Leaf*)

DRAWING CHECKLIST
CLARITY:
large
clean/ smooth
no shading
ACCURACY
specimen
proportion
LABELLING
parallel
accurate lower case
justified
annotation
magnificati
title & view
TOTAL

CALCULATION OF MAGNIFICATION:

$$\frac{\text{magnification of drawing}}{} = \frac{\text{size of drawing}}{\text{size of leaf}}$$

$$= \underline{\qquad} / \underline{\qquad}$$

$$= \underline{X \qquad}$$

TITLE: _____

REFLECTION – *This lab gave you a better understanding or appreciation of ...?*

_____

_____

NOTES: DESCRIBING LEAVES – EXTERNAL FEATURES

The leaf is a flattened, lateral outgrowth of the stem in the branch. It develops from a node and has a bud in its axil. It is normally green and makes food - carbohydrates for the whole plant through the process of photosynthesis. Parts of the leaf are described below in table 1 and table 2 below:

TABLE 1 – DESCRIPTIVE FEATURES OF A LEAF.

LEAF FEATURES - DESCRIPTIONS

1. Shape	lance elliptic oval heart round
2. Apex or tip (pointed or rounded)	Acute Acuminate Cuspidate Rounded Long tapering
3. Margins or edges (smooth, serrated)	Entire Serrate/ Toothed Undulate Doubly serrate Lobed Crenulate
4. Bases	Cuneate Rounded Truncate Cordate Oblique Auriculate
5. Lamina	• Smooth • Hairy or velvety • Waxy • Thick • Thin
6. Colour	Dark/light green; variegated (green and white/ mixture of colours); patterned.
7. Venation (veins)	• network of veins or branched veins (found in dicot plants) • veins run parallel to mid-rib (found in monocot plants) • prominent mid-rib
8. Petiole	Long, short, thick, slender, absent

TABLE 2. WITH ANNOTATIONS FOR THE DRAWING OF A HIBISCUS LEAF.

Part	Description	Function
Lamina	- Green and broad - Thin - flat - waxy	• contains numerous chloroplasts for photosynthesis • allows light to go to the mesophyll to make food; • allows CO_2 to reach all cells rapidly • reduces water loss by evaporation
Petiole	See table 1 above	• holds the lamina away from the stem.
Mid rib/ main vein	contains vascular tissues	• supply water and minerals as well as transports food. • hold leaf upright.
Lateral veins	branch off the main vein	• supply all cells with water take food away from the leaf

BIOLOGY CSEC MARK SCHEME FOR DRAWING (DR)

CRITERIA (DRAWING – DR)			Max. Marks	Teacher Marks
C- CLARITY	l c s	- Large drawing (½ page or larger) (1) - Clean, smooth, thin, continuous lines (1) - No shading or unnecessary details (1)	3	
A- ACCURACY	s p	- Looks like specimen (1) - Reasonable proportions (1)	2	
L-LABELLING and LABEL LINES	p a j	- Straight, no arrow head, parallel (1) - Accurate label, lowercase, letters not joined (1) - Annotations – at least 3 (1) - Justified labels (start at some point)	3	
	m	- Magnification calculation correct (1)	1	
	t	- Title – at bottom, in CAPITALS, underlined - View of specimen stated	1	
TOTAL			10	

Additional task: Compare the external features of different types of dicotyledonous plant leaves within the school compound such as ixora, vinca/periwinkle, Pride of Barbados.

NAME:_____**CLASS:**_____   **SKILL:** AI/MM

TOPIC: PHOTOSYNTHESIS   SYLLABUS OBJECTIVE: B 2.4

6.   INVESTIGATING HOW LIGHT AFFECTS PHOTOSYNTHESIS

AIM:_____

_____

$$6\ CO_2\ + 6\ H_2O\ \xrightarrow[chlorophyll]{sunlight}\ C_6H_{12}O_6\ +\ 6O_2$$

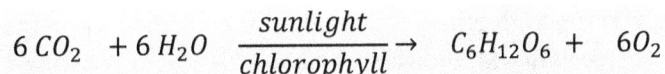

What will happen if there were no sunlight? Would we be able to get food? Can green plants make food without sunlight? The process of photosynthesis is affected light intensity, presence of chlorophyll, water availability, carbon dioxide concentration and even temperature. This experiment will examine how just one factor – light – affects photosynthesis. How can we determine if photosynthesis is occurring? Identification of either starch (food) or oxygen bubbles, the products of photosynthesis.

APPARATUS AND MATERIALS:

- Boiling tube
- Petri dish
- White tile
- Water bath – large beaker with water
- Bunsen burner and Tripod stand
- Forceps/tweezers
- Test tube rack
- Test tube holder

- Foil
- Paperclips
- Green leaved potted plant (non-waxy, small leaves)
- Dark room/ cupboard/ black bag
- Iodine solution
- Ethanol
- Labels

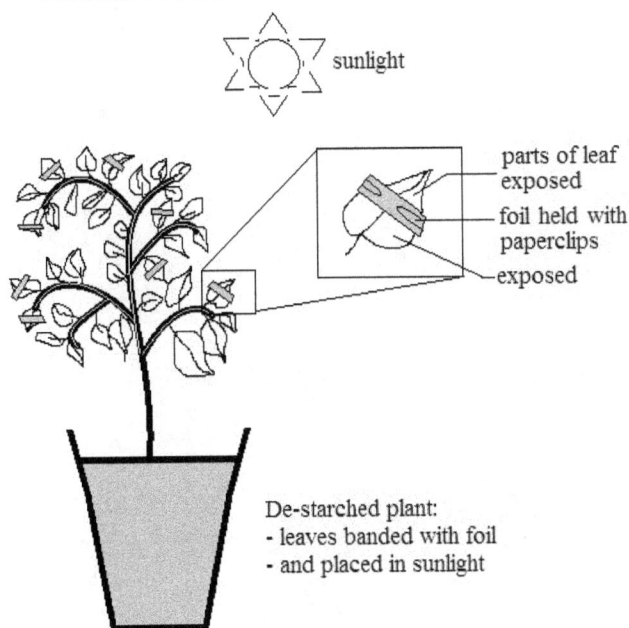

sunlight

parts of leaf exposed
foil held with paperclips
exposed

De-starched plant:
- leaves banded with foil
- and placed in sunlight

SET-UP OF PLANT WITH THE FOIL AND PAPER-CLIP.

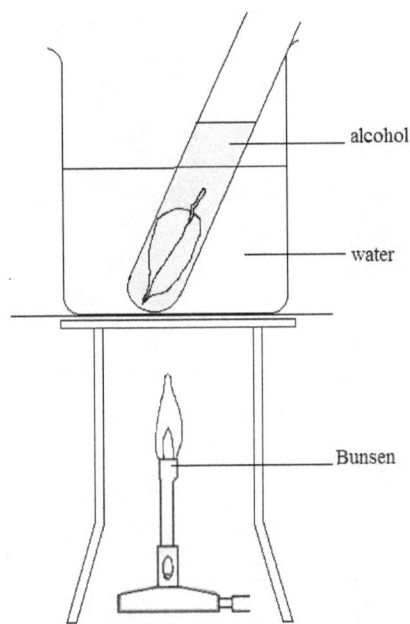

alcohol

water

Bunsen

SET-UP OF THE WATER-BATH USED TO DECOLOURIZE THE LEAF BEFORE THE STARCH TEST.

INSTRUCTIONS:

1. De-starch a small green-leaved plant by placing it in a dark room/cupboard for 24 hours.
2. Select 1 leaf and cover part with a strip of foil paper. Use the paperclip to hold the foil in place.
3. Place the plant in sunlight for at least 3 hours.
4. Pick the leaf from the plant, remove the foil and paperclip.
5. Set up a water bath.
6. Boil the leaf for 10 seconds in the water to remove the waxy cuticle, carefully remove it with the forceps and place it in the boiling tube.
7. Cover the leaf completely with ethanol; then place the boiling tube in the water bath until the leaf loses all its green colour. (Be careful: the ethanol is very flammable!)
8. Rinse the leaf and place it in a Petri dish over a white tile. Add iodine solution to completely cover the leaf.
9. Observe and record the patterns of colour that develop on the leaf.

Rewrite your method in the past tense in the space below or on a separate page.

METHOD:

RESULTS:

C - Draw the leaf before boiling. ANNOTATE the colour of the leaf and where the foil was placed.

TITLE:_____

Draw and annotate the colour and appearance of the leaf <u>AFTER THE STARCH TEST</u> (when iodine was added).

TITLE:_____

DISCUSSION

LIMITATIONS _____

_____

_____

_____

_____

PRECAUTIONS _____

_____

_____

_____

SOURCES OF ERROR _____

_____

_____

_____

CONCLUSION *(What condition is needed for photosynthesis? Why is that so? Restate the results.)*

_____

_____

_____

_____

REFLECTION *(What do you have a better understanding/ appreciation for with respect to photosynthesis? / What was most significant, meaningful or useful?/ What will your next step be?)*

_____

_____

_____

_____

CSEC Sample mark scheme for AI skill – To see if light is needed for photosynthesis

AI GENERAL CRITERIA			SPECIFIC CRITERIA – To see if light is needed for photosynthesis.	MAX MARK	T. Mk.
B-Background	d s	• Define key related terms (1) • Statement of relevant theory (1)	Photosynthesis defined correctly Functions of chlorophyll AND sunlight	2	
E - Explanation	t c u m	- Trends and patterns identified (1) - Compare actual results with expected results (1) - Use data to support explanations (1) - Modification/ improvement to existing method (1)	-Which parts became blue black or yellow brown. -Expectations on blue-black area and brown areas compared to actual results. -Reference made to results drawing or annotations of part s of the leaf. -Improve method – decolourize more, softer leaves, etc. – must be EXPLAINED	4	
LSP - Limitations Sources of error/ Precautions		- Identify at least 2 limitations **with explanations** - Identify at least 2 precautions/ sources of error **with explanations**	-Limitations – cloudy skies, improper destarching, not enough sun exposure, etc. -Precautions – see hints -Sources of Error – if results were inconclusive – decolourize more, use softer/ less waxy leaf.	2	
C-Conclusion	s r	- Statement - Related to aim	-_____ is needed for photosynthesis because _____ -From the results/ due to sources of error, it could not be determined if light is…..	2	
TOTAL				10	

SUGGESTED MARK SCHEME – MM CRITERIA	Marks	Teacher Mk
• Leaf covered securely with foil paper and paper clip (does not raise)	1	
• Leaf boiled for 10 seconds in the boiling water to remove wax	1	
• Forceps used to handle the leaf at all times throughout the exp't.	1	

• Water bath correctly set up (1 mark each)
   o 2/3 fill beaker with tap water
   o Blue flame to allow water to boil
   o Switch off flame/ Bunsen burner when boiling tube added
	3	

• Leaf completely covered in ethanol • Leaf completely decolourized before moving on to the next step	2	
• Mouth of boiling tube facing away from students and books	1	
• Test tube holder used to remove boiling tube from water bath o Clamped under the rim of the boiling tube	1	
• Leaf covered with enough iodine solution for blue-black colour to appear	1	
• Work station/ lab bench/ table top cleaned and stools put away	2	
TOTAL	13	

DISCUSSION QUESTIONS *(Answer ALL)*

1. Define the term photosynthesis giving the word and chemical equations (on separate lines).
2. Explain the functions of chlorophyll and sunlight – necessary conditions for photosynthesis.

3. Before the experiment, <u>what colour results were EXPECTED?</u> Compare with the recorded results.
4. Explain what happened in the blue-black parts of the leaf. Explain the brown parts of the leaf.

5. In the experiment, explain why
    a. The plant was de-starched before placing it into the sunlight.
    b. The leaf was decolourized by boiling in ethanol before iodine solution was added.
    c. Iodine solution was used.

$$6\,CO_2 \;+\; 6\,H_2O \xrightarrow[chlorophyll]{sunlight} C_6H_{12}O_6 \;+\; 6O_2$$

RAW MATERIALS   ESSENTIAL REQUIREMENTS   PRODUCTS

6. Identify the control in this experiment. If there was none, what would it have been?
7. How can the method be improved for accurate results?

8. What limitations or factors were difficult to control?
9. What precautions were taken? (de-starching; completely; covering part of the leaf with foil tightly; careful use of flammable ethanol, decolourizing completely)
10. Identify any sources of error.

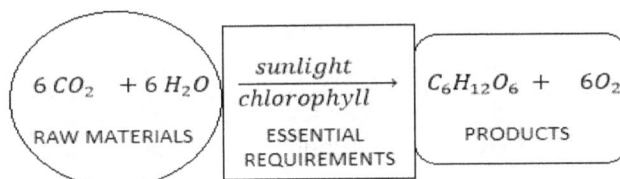

NOTES - PHOTOSYNTHESIS

There are two stages in the photosynthesis process – 1. **Sunlight splits the water** molecule to release hydrogen and oxygen (waste gas). 2. The **hydrogen then combines with carbon dioxide** to form carbohydrates such as glucose. Glucose is the energy source for cellular processes. Individual glucose molecules combine to form complex carbohydrates- starch - stored in the leaf, and, cellulose - makes cell walls.

In order to test for starch, a fresh green leaf must be boiled in water to remove the waxy cuticle. This is so the iodine solution can penetrate the cells. The leaf also has to be de-colourized (all the chlorophyll molecules removed with ethanol) so that the green colour will not hide the colour of the iodine.

PHOTOSYNTHESIS

SUN

Light Energy
Splits water molecule to start photosynthesis in light dependent stage

Oxygen gas

Leaves – make glucose (food) stored as starch

Chlorophyll
- In chloroplasts of mesophyll cells
- Traps all wavelengths of light except green

Carbon dioxide
- Diffuse through stomata
- Then palisade mesophyll
- Fixed to hydrogen in light independent stage

Oxygen gas
-exit leaves through stomata

Water
- Absorbed through roots
- Carried to leaves via xylem in stems

Roots - Store starch (food) transported after photosynthesis

Iodine solution causes starch to turn blue-black. The leaf is tested for starch molecules, stored from photosynthesis. It is expected that the parts of the leaf exposed to sunlight will photosynthesize and form starch (giving a positive blue black colour with iodine). However the parts of the leaf under the foil are expected to give a negative yellow brown colour with iodine solution. If these results are obtained, this shows that (i) sunlight is necessary for the process of photosynthesis, and (ii) leaves make starch as food by photosynthesis.

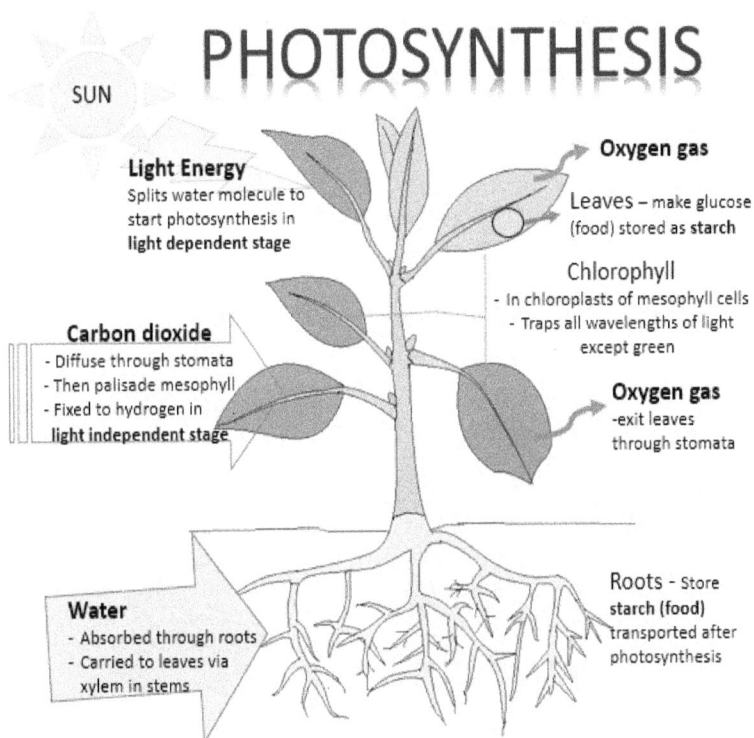

NAME:_____**CLASS:**_____   **SKILL: ORR/MM**

TOPIC: PHOTOSYNTHESIS   SYLLABUS OBJECTIVE: B 2.4

7.   INVESTIGATING HOW CHLOROPHYLL AFFECTS PHOTOSYNTHESIS

AIM TO TEST IF CHLOROPHYLL IS NEEDED FOR PHOTOSYNTHESIS

To test if chlorophyll is needed for photosynthesis, a variegated leaved plant is needed. Instead of covering parts of green leaves with foil, a variegated leaf is ready to test after de-starching and then sun exposure. Variegated leaves are partly green and partly white. The green part of such a leaf contains chlorophyll but the white part of such a leaf does not contain chlorophyll. A small soft variegated leaf may be used as found on the Chinese violet *Asystasia gangetica variegata*.

APPARATUS AND MATERIALS:

Same as previous lab, except use a variegated leaved plant. Foil and paper clip not required.

INSTRUCTIONS

Same as previous lab (pg 70), leaving out steps 2 and 4.

RESULTS:

Draw and annotate the colour and appearance of the leaf BEFORE AND AFTER THE STARCH TEST (when iodine was added).

TITLE:_____

NAME:_____ CLASS:_____ SKILL: **MM/ORR**

TOPIC: NUTRITION   SYLLABUS OBJECTIVE: B 2.6, 2.11

8.   FOOD TESTS TO IDENTIFY CARBOHYDRATES - REDUCING AND NON-REDUCING SUGARS

AIM:_____

_____

Do doubles constitute a balanced meal? Why is milk so nutritious? The food you eat contains some combination of biomolecules such as carbohydrates (sugars and starch), fats and proteins. This experiment allows students to become familiar with both the positive and negative colour results when identifying biomolecules such as carbohydrate sugars (reducing and non-reducing). A food sample can also be tested for these biomolecules it may contain.

APPARATUS AND MATERIALS:

- 6 Test tubes in racks
- 2 Test tube holders
- 2 Measuring cylinders
- 2 Syringes
- Large beaker
- Bunsen burner
- Gauze and tripod stand

- 2 Stirring rods
- Labels
- Stop clock
- Glucose solution (reducing sugar)
- Sucrose solution (non-reducing sugar)

- Distilled water
- Benedict's solution
- Hydrochloric acid (HCl)
- Sodium hydroxide (NaOH)
- *Food sample _____ (optional)

DIAGRAM:

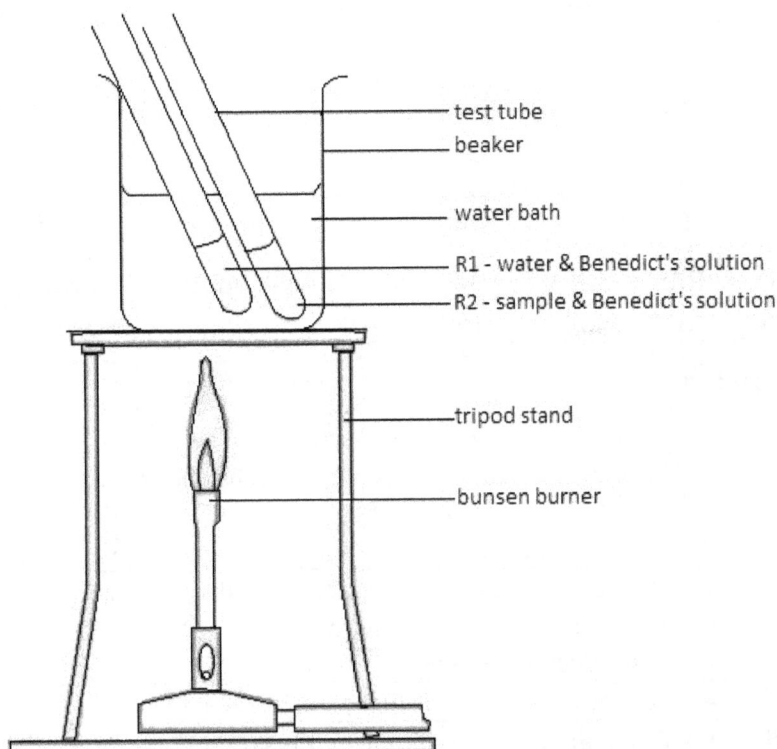

- test tube
- beaker
- water bath
- R1 - water & Benedict's solution
- R2 - sample & Benedict's solution
- tripod stand
- bunsen burner

DIAGRAM SHOWING THE SET UP OF WATER BATH FOR REDUCING SUGAR TESTS.

INSTRUCTIONS/METHOD:

NOTE: Solid food samples must be crushed and liquefied with distilled water before testing.

(A) BENEDICT'S TEST (REDUCING SUGAR TEST):

1. Label 2 clean test tubes – R1 and R2. Add 1cm³ distilled water to R1. Add 1cm³ glucose to R2.
2. In **EACH** test tube add 2 cm³ of Benedict's solution. Observe and record the initial colours.
3. Place both test tubes into a boiling water bath for 2 minutes.
4. Observe and record the <u>progress to final colours</u> (on heating) of each test tube in the table.

(B) NON-REDUCING SUGAR TEST:

1. Label 2 clean test tubes – N1 and N2. Add 1cm³ distilled water to N1. Add 1cm³ sucrose to N2.
2. In **EACH** test tube add 2 cm³ hydrochloric acid (HCl) and boil **each** tube in a water bath for exactly 1 minute.
3. After removing, neutralize **each** tube with 2cm³ sodium hydroxide (NaOH) until fizzing stops.
4. Carry out the Benedict's test as above (A -steps 2 – 4).

<u>TABLE 1 SHOWING CONTENTS OF EACH TEST TUBE AND TEST DONE.</u>

Name of Test	Substance tested	Main reagents added/ treatment
Benedict's test/ **Reducing sugar test**	R1 - water R2 – glucose solution R3 – food sample _____	- Benedict's solution - Heat
Non-reducing sugar test	N1 - water N2 – sucrose solution N3 – food sample _____	- HCl + heating then NaOH - Benedict's solution + heat

SUGGESTED MARKSCHEME – MANIPULATION AND MEASUREMENT (MM)

SAMPLE MARK SCHEME – MM CRITERIA		Marks	Teacher Mk
Apparatus	**Steps**		
SYRINGE	• Plunger fully depressed at start	1	
	• No air bubbles when filled	1	
	• Read meniscus at point of upper rubber ring	1	
	• Accurate interpretation of scale	1	
	• Rinse between uses	1	
STOP CLOCK	• Ensure spring is wound/ clock is working	1	
	• Start at moment tubes placed in water bath	1	
	• Reset clock when test tubes removed	1	
TEST TUBES	• Face away from others while in water bath	1	
TEST TUBE HOLDER	• Clamp under rim to remove hot test tubes	1	
	TOTAL	**10**	

RESULTS: *(In the table below, record the observations, i.e. <u>initial and final</u> colours of mixtures. Based on the colours recorded make an <u>inference</u> for each substance tested, e.g. "Positive for reducing sugars" or "negative for reducing sugars".*

TABLE TITLE:_____

Name of test	Substance tested	Observations – colours of mixtures		Inference
		Initial (before heating)	Final	
Benedict's test/ Reducing sugar test	R1 – water		*	
	R2 – glucose solution		*	
	R3 -		*	
Non-reducing sugar test	N1 – water		*	
	N2 – sucrose solution		*	
	N3 -		*	

*Note progression of colours as they change from the initial blue through to yellow, green, orange, red and finally brick red colour.

Under the column *Final Colour* - **DO NOT record "no colour change"**. Instead write the actual colour seen e.g. clear or colourless or blue OR "blue → green → orange → red".

DISCUSSION

_____

_____

_____

_____

_____

_____

_____

_____

_____

_____

_____

_____

_____

_____

_____

_____

_____

LIMITATIONS

_____

_____

PRECAUTIONS

_____

SOURCES OF ERROR

_____

CONCLUSION (*What was found out in this experiment about food tests for the biomolecules - reducing sugar and non-reducing sugar? Make a statement on the methods for these tests.*

_____

_____

_____

REFLECTION (*How can food test experiments and results be relevant to your everyday life?*)

_____

_____

_____

DISCUSSION QUESTIONS:

1. What are biomolecules and why are they important?
2. State the function of each of the two biomolecules, reducing sugars and non-reducing sugars in the body.
3. What were the expected results? Identify the control in each case.
4. Based on the results, what were the positive colours for each biomolecule?

5. What precautions were taken in the experiment? (Hint – washing the measuring cylinders between each test conducted/ different reagent used; placing the test tubes into the water bath at the same time.)
6. Identify any sources of error in achieving the correct expected results.
7. What were possible limitations? (Hint: if food sample is already coloured before testing then...)

NOTES – FOOD TESTS/ BIOMOLECULES PRESENT IN FOOD

Food tests are a simple way to **identify the biomolecules/ nutrients present** in samples of food. Biomolecules may be complex carbohydrates like starch or simple sugars like glucose or sucrose, or they may be proteins or lipids. Food tests use the basis of **positive colour results to simply <u>identify the presence</u>** of a particular biomolecule. However, it is difficult to determine the quantities or proportions of a particular nutrient present in a food. In general, the foods tested must be in liquid form. Solid food samples must be crushed and liquefied before testing so the reagents can work faster.

Humans eat food in order to get the **nutrients necessary to carry out life** functions. Most food contain a combination of biomolecules or nutrients, for example, bread contains starch and some reducing sugars; milk contains sugars, proteins and lipids. Each of the **six food groups** – 1. Staples, 2. Legumes, 3. Food from animals, 4. Vegetables, 5. Fruits and 6. Fats and oils – provides some or all of the nutrients. **A balanced diet** provides the correct proportions of each food group that will give the necessary biomolecules/nutrients to survive.

Carbohydrates in general supply energy to the body. Some complex carbohydrates like cellulose provide the body with roughage or fibre; this is not tested for in the food tests. **Starch** another complex carbohydrate must be broken down using enzymes in the gut, into simple sugars. These simple sugars (reducing sugars and non-reducing sugars) provide energy for cells during respiration. Reducing sugars like **glucose** or fructose are so called because they **reduce the blue copper ions (Cu^{2+}) to red (Cu^+).** Heating speeds up reduction. In general, with increasing reducing sugars present the more the red colours are observed and less time it takes to develop on heating. However, the **colour changes with reducing sugars go from blue to green to yellow then orange and finally (brick) red.**

If the Benedict's test was done on sucrose, it would give a negative blue colour result, even after heating. In order to identify **non-reducing sugars like sucrose,** the sucrose has to undergo **acid hydrolysis first** (using a dilute acid (HCl) to split the disaccharide sucrose into two monosaccharides - glucose and fructose). The acid must then be neutralized by NaOH or other alkaline solution. Only then will the Benedict's test give the colour changes described above.

TABLE IDENTIFYING THE FOOD NUTRIENTS/ BIOMOLECULES PRESENT IN DIFFERENT FOOD GROUPS

Food groups	Amounts of Food nutrients				
	Carbohydrates	Proteins	Fats	Vitamins and Minerals	Fibre
Staples e.g. bread, roti, wheat, cereal, potato, rice, flour	✓✓✓	✓	✓	✓✓	✓
Legumes and nuts e.g. pigeon peas, channa, peanuts	✓✓✓	✓✓✓	✓	✓	✓
Animal foods e.g. chicken, beef, lamb, fish, milk	✗	✓✓✓	✓✓	✓✓	✗
Fats and Oils e.g. butter, coconut, avocado, vegetable oil	✗	✗	✓✓✓	✓✓	✗
Fruits e.g. oranges, guava, mango, pawpaw	✓✓	✗	✗	✓✓✓	✓✓
Vegetables e.g. carrots, christophene, broccoli, okro	✓	✗	✗	✓✓✓	✓✓✓

✗ - none present   ✓, ✓✓, ✓✓✓ - increasing amounts of nutrients

NAME:_____**CLASS:**_____ **SKILL: MM/ORR**

TOPIC: NUTRITION   SYLLABUS OBJECTIVE: B 2.6, 2.11

9.   FOOD TESTS TO IDENTIFY STARCH, PROTEIN AND LIPIDS

AIM:_____

_____

Do doubles constitute a balanced meal? Why is milk so nutritious? The food you eat contains some combination of biomolecules such as carbohydrates (sugars and starch), fats and proteins. This experiment allows students to become familiar with both the positive and negative colour results when identifying biomolecules such as carbohydrate sugars (reducing and non-reducing). A food sample can also be tested for these biomolecules it may contain.

APPARATUS AND MATERIALS:

- 9 Test tubes in racks
- Test tube holder
- 2 Measuring cylinders
- 2 Syringes
- Labels

- Distilled water
- Starch solution
- Oil (vegetable)
- Protein solution (albumen)

- 5% Copper sulphate ($CuSO_4$) solution
- Iodine solution
- Ethanol
- *Food sample _____ (optional)

DIAGRAM:

(Include a drawing of the set-up of the apparatus)

INSTRUCTIONS/METHOD:

NOTE: Solid food samples must be crushed and liquefied before testing; 1cm³ of the mixture should be tested..

(A) STARCH TEST (IODINE TEST)

1. Using cleaned test tubes – S1 and S2 – add 1cm³ distilled water to S1. Add 1cm³ starch to S2.
2. Add 3 drops of iodine solution to **EACH** test tube.
3. Observe and record the initial and final colours of each test tube in an appropriate table.

(B) PROTEIN/ BIURET TEST

1. Using cleaned test tubes – P1 and P2 – add 1cm³ distilled water to P1. Add 1cm³ protein to P2.
2. To **EACH** test tube add 1 cm³ of sodium hydroxide (NaOH) followed by 4 drops of 5% Copper Sulphate solution (CuSO₄), shake gently to mix each test tube.
3. Observe and record the initial and final colours of each test tube in an appropriate table.

(C) EMULSION TEST FOR LIPIDS

1. Using cleaned test tubes – L1 and L2 – add 1cm³ distilled water to L1. Add 1cm³ oil to L2.
2. To each test tube add 1 cm³ of ethanol, and shake vigorously.
3. Pour cold distilled water into each test tube.
4. Observe and record the initial and final colours of each test tube in an appropriate table.

ALTERNATIVE TEST FOR LIPIDS – GREASE SPOT TEST

The emulsion test for lipids is not the only possible test for lipids. A simpler method involves rubbing a sample of food (solids) on a piece of filter paper or other white paper. After drying (water evaporates), a translucent spot should remain on the paper where the lipids touched it.

TABLE 1 SHOWING CONTENTS OF EACH TEST TUBE AND TEST DONE.

Name of Test	Substance tested	Main reagents added/ treatment
Starch/ Iodine test	S1 - water S2 – starch solution S3 – food sample _____	- Iodine solution - (***no heating***)
Protein/ Biuret test	P1 - water P2 – protein solution P3 – food sample _____	- NaOH or KOH - 5% CuSO₄ - (***no heating***)
Emulsion test	L1 - water L2 – oil L3 food sample _____	- Ethanol + shaking - Cold water

RESULTS: *In the table below, record the observations, i.e. <u>initial and final</u> colours of mixtures. Based on the colours recorded make an <u>inference</u> for each substance tested, e.g. "Positive for starch" or "Negative for starch". Identify the control substance in each test.*

TABLE TITLE:_____

Name of test	Substance tested	Observations – colours of mixtures		Inference
		Initial	Final	
Starch/ Iodine test	S1 – water			
	S2 – starch solution			
	S3 -			
Protein/ Biuret test	P1 – water			
	P2 – protein solution			
	P3 -			
Emulsion test	L1 – water			
	L2 – oil			
	L3 -			

Under the column *Final Colour* - **DO NOT record "no colour change"**. Instead, write the actual colour seen e.g. blue, orange, clear or colourless.

DISCUSSION

_____

_____

_____

_____

_____

_____

_____

_____

_____

_____

_____

_____

_____

_____

_____

_____

_____

LIMITATIONS

_____

_____

PRECAUTIONS

_____

SOURCES OF ERROR

_____

CONCLUSION _(What was found out in this experiment about food tests for the biomolecules – starch, protein and lipids/oil? Make a statement on the methods for these tests._

_____

_____

_____

REFLECTION _(How can food test experiments and results be relevant to your everyday life?)_

_____

_____

_____

DISCUSSION QUESTIONS:

1. What are biomolecules and why are they important?
2. State the function of each of the biomolecules in the body, starch, proteins, and lipids.

3. What were the expected results? Identify the control in each case.
4. Based on the results, what were the positive colours for each biomolecule?

5. What precautions were taken in the experiment? (Hint – washing the measuring cylinders between each test conducted/ different reagent used; placing the test tubes into the water bath at the same time.)
6. Identify any sources of error in achieving the correct expected results.
7. What were possible limitations? (Hint: If food sample is already coloured before testing then…)

NOTES – FOOD TESTS/ BIOMOLECULES PRESENT IN FOOD

Food tests are a simple way to **identify the biomolecules/ nutrients present** in samples of food. Biomolecules may be complex carbohydrates like starch or simple sugars like glucose or sucrose, or they may be proteins or lipids. Food tests use the basis of **positive colour results to simply <u>identify the presence</u>** of a particular biomolecule. However, it is difficult to determine the quantities or proportions of a particular nutrient present in a food. In general, the foods tested must be in liquid form. Solid food samples must be crushed and liquefied before testing so the reagents can work faster.

Humans eat food in order to get the **nutrients necessary to carry out life** functions. Most food contain a combination of biomolecules or nutrients, for example, bread contains starch and some reducing sugars; milk contains sugars, proteins and lipids. Each of the **six food groups** – 1. Staples, 2. Legumes, 3. Food from animals, 4. Vegetables, 5. Fruits and 6. Fats and oils – provides some or all of the nutrients. **A balanced diet** provides the correct proportions of each food group that will give the necessary biomolecules/nutrients to survive.

In general carbohydrates supply energy to the body. Some complex carbohydrates like cellulose provide the body with roughage or fibre; this is not tested for in the food tests. **Starch, another complex carbohydrate, causes yellow iodine to become blue-black.** Starch is a storage carbohydrate for plants, but in the human body it has to be broken down with amylase enzyme into simple sugars to make energy available via respiration. Thus energy provided by the breakdown of starch takes some time to be supplied.

Proteins are organic molecules needed for growth, repair, and formation of enzymes, hormones and antibodies (for immunity). Proteins are generally sourced from meat and dairy products. Plant sources of proteins include beans (chickpeas/ channa) and nuts (peanuts and cashews). The **protein test gives a positive purple/ mauve colour** as the peptide bonds that hold amino acids together interact with the blue copper ions. If the solution remains light blue, then no proteins are present.

Lipids can be either fats – solid at room temperature, or, oils – liquid at room temperature. Lipids can be a **source of energy** but are usually stored under the skin and around major organs and provide insulation. The **emulsion test** identifies oils present when the alcohol dissolves the oil, forming **tiny globules of fat that look milky white** when water is added, **i.e. an emulsion.**

Method A – Starch Test; Method B – Protein Test and Method C – Emulsion Test

SAMPLE MARK SCHEME – MM CRITERIA			Maximum Marks	Teacher Mark
Apparatus		**Steps**		
HANDLING REAGENTS/ BOTTLES	•	Read label before use	1	
	•	Pour away from the label	1	
	•	Stopper placed upside down	1	
	•	Replace stopper after pouring	1	
	•	Reagents NOT returned to stock bottle if too much is poured out	1	
DROPPER BOTTLE	•	Same size drops when using dropper	1	
CLEANLINESS	•	Clean test tubes between uses	1	
	•	Clean up of work station	1	
	•	Putting away stools/ apparatus/etc	1	
PREPARATION/ EXECUTION	•	Read the experiment and followed the steps correctly	1	
		TOTAL	**10**	

SAMPLE MARK SCHEME - ORR GENERAL CRITERIA					Max. Mark	T. Mk.
O – Observations	s o c/d	• Significant changes noted – correct final colours for all test substances (1) • Original and final conditions – recorded for all tests (1) • Control noted OR diagram (1)			3	
RTG – Recording	t u e	**TABLE** • Title – above, in capitals, underlined (1) • Column & row headings(with units) (1) • Enclosed and neat (1)	t u p	**GRAPH** • Title below, capitals, underlined (1) • Both axes labelled with units (1) • Accurate plots (1)	3	
R – Reporting	a r g s	• Aim in capital letters (1) • Reflection appropriate (1) • Acceptable language and expression – subject-verb agreement/ grammar (1) • Spelling correct throughout with 0-3 errors (1) *Note: > 3 grammatical errors = 0 marks. AND > 3 spelling errors = 0 marks*			4	
TOTAL					**10**	

NAME:_____**CLASS:**_____ **SKILL: AI____**

TOPIC: RESPIRATION   SYLLABUS OBJECTIVE: B 3.2

10.   INVESTIGATING THE PRODUCTION OF CARBON DIOXIDE FROM ANAEROBIC RESPIRATION
       OF YEAST

AIM:_____--

_____

All living things carry out respiration to provide their cells with energy. How do we know? When energy
is formed, a waste gas carbon dioxide is made; and this can be identified in the lab with limewater which
becomes cloudy. Yeast is a type of fungus that respires both aerobically and anaerobically. Anaerobic
respiration of yeast produces carbon dioxide and alcohol, and it will be investigated in this experiment
by identifying the carbon dioxide gas formed.

APPARATUS AND MATERIALS:

- 2 Boiling tubes
- 2 Test tubes
- Test tube rack
- 2 Measuring cylinders
- 2 Delivery tubes
- 4 rubber bungs (2 large and 2 small)
- Distilled water
- Beaker

- Yeast/ Yeast suspension
- Glucose sugar/ 10% glucose solution
- Oil
- Labels
- Limewater ($Ca(OH)_2$)
- Water bath between 35-40°C

DIAGRAM

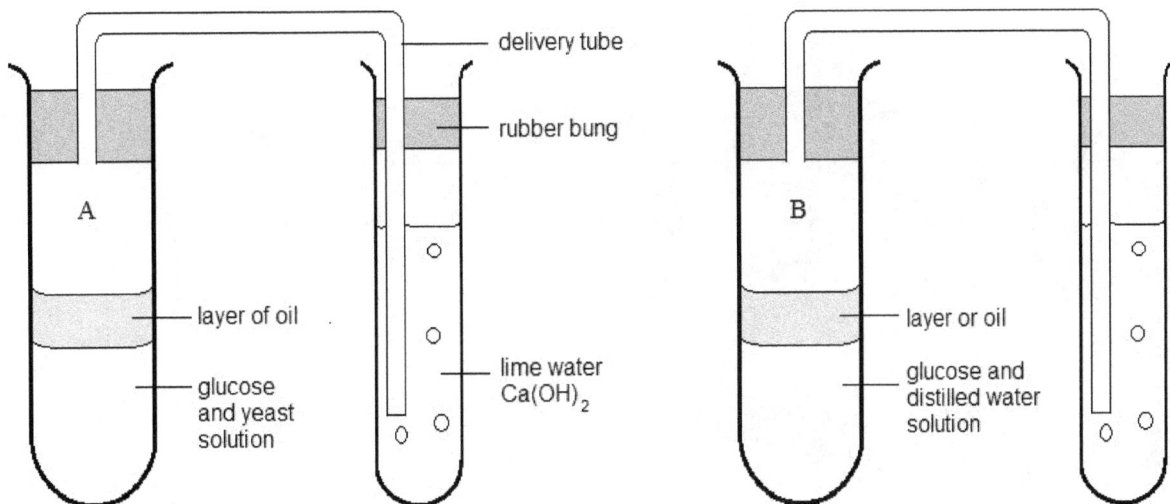

DRAWING SHOWING SET UP OF APPARATUS TO INVESTIGATE ANAEROBIC RESPIRATION OF YEAST.

INSTRUCTIONS:

1. Boil 40 cm^3 of water in a beaker to remove the dissolved air.
2. Dissolve 1spatula of glucose sugar in the warm water, then leave the mixture to cool completely.
3. Set up 2 test tubes with approximately 10cm^3 limewater.
4. Label 2 boiling tubes A and B respectively.
5. Add 5cm^3 of the cooled glucose solution to each boiling tube.
6. Carefully add a 5cm^3 of yeast suspension to boiling tube A. For boiling tube B, add 5 cm^3 distilled water.
7. Carefully pour oil slowly down the sides of boiling tubes A and B so it forms an oil layer over the mixtures in each boiling tube as in the diagram.
8. Cover each boiling tube with the large rubber bung and the short end of the delivery tube. Place the longer end of the delivery tube in a test tube with lime water (see diagram of set up).
9. Construct a table to note the initial appearances of the contents of the boiling tubes and the test tubes in each set up.
10. Leave both pieces of apparatus in a warm place for half an hour or a water bath at 35-40°C.
11. Observe and record what happens to the mixtures and the lime water in each tube (boiling tube and test tube) in each set-up.

Rewrite your method in the past tense in the space below or on a separate page.

METHOD:

RESULTS:

Arrange the worksheet below to record the results. Use either a table or annotated diagrams for each set-up used to identify carbon dioxide produced from anaerobic respiration of yeast.

TITLE: _____

DISCUSSION

LIMITATIONS _____

_____

_____

PRECAUTIONS _____

_____

_____

SOURCES OF ERROR _____

_____

_____

_____

CONCLUSION *(Explain how anaerobic respiration of yeast was identified in this experiment)*

_____

_____

_____

_____

_____

_____

REFLECTION *(How is the production of carbon dioxide by anaerobic respiration of yeast relevant to your everyday life? What do you have a better understanding about, having done this lab?)*

_____

_____

_____

_____

_____

_____

_____

CSEC Sample mark scheme for AI skill – To investigate the anaerobic respiration of yeast

AI GENERAL CRITERIA			SPECIFIC CRITERIA Anaerobic respiration of yeast	MAX MARK	T. Mk.
B- Background	d	• Define key related terms (1)	- Anaerobic respiration defined/ equation	2	
	s	• Statement of relevant theory (1)	- How yeast used to make bread/ alcohol		
E - Explanation	t	- Trends and patterns identified (1)	- How presence of carbon dioxide gas is identified/ where alcohol formed	4	
	c				
		- Compare actual results with expected results (1)	- Expectations stated and compared for both A and B set-ups. Identify any sources of error here.		
	u	-			
		- Use data to support explanations (1)	- Specific reference made to the observations recorded for A (boiling tube and test tube) and B (both tubes)		
	m				
		- Modification/ improvement to existing method (1)	- Modification suitable/ temperature explained how it improves results.		
LSP - Limitations Sources of error/ Precautions		- Identify at least 2 limitations **with explanations**	- Two limitations explained correctly	2	
		- Identify at least 2 precautions/ sources of error **with explanations**	- Precautions taken for set-up to remain anaerobic or set-up to equilibrate. - Restate source of error and explain.		
C- Conclusion	s r	- Statement - Related to aim	How limewater indicated that carbon dioxide was formed due to anaerobic respiration of yeast….	2	
TOTAL				10	

DISCUSSION QUESTIONS

1. What is respiration and why is it important?
2. How is anaerobic respiration of yeast useful/ important in everyday life and in industry?

3. Why is anaerobic respiration of yeast investigated and not aerobic respiration?
4. What two precautions were taken in the method to ensure conditions were anaerobic? (Hint: Boiling the water for the glucose solution and the layer of oil)
5. Why was the glucose solution cooled before yeast was added?

6. Identify the control set-up in this experiment and say how it is useful. (Note: in the control set up, limewater should remain clear and transparent throughout the experiment)
7. What observations were expected for set-up A and for B? Do the actual results confirm this?
8. Based on the results, explain what happened to the lime water in set-up A and B? Write a chemical equation to show how limewater turned cloudy.
9. What new substance would be expected in the boiling tube with the live yeast at the end of the experiment? How can it be identified (Hint: scent/ igniting)

10. Explain how the method can be improved. (Hint: Constant temperature with a water bath)
11. Identify sources of error that may have affected the results (leaking rubber bungs, not enough time for yeast to multiply/inactive yeast cells; expired calcium hydroxide).
12.
13. What limitations or factors were difficult to control?

NOTES - RESPIRATION

Yeast belongs in the Fungi Kingdom and produces respiration enzymes that breakdown the substrate glucose to obtain energy. Anaerobic respiration is the **incomplete breakdown of carbohydrates** in the absence of oxygen. **Less energy** is produced from anaerobic respiration.

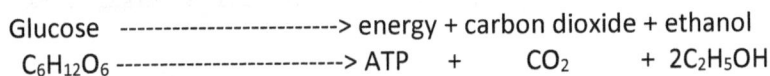

Glucose ------------------------> energy + carbon dioxide + ethanol
$C_6H_{12}O_6$ ------------------------> ATP   + CO_2 + $2C_2H_5OH$

When yeast undergoes anaerobic respiration to break down glucose, the products ethanol and carbon dioxide, are wastes for the yeast cells. This **anaerobic respiration of yeast is called alcoholic fermentation** since the alcohol produced is very important in commercial industry. Different alcohols are produced by fermenting different sources of sugar with the yeast cells. Humans also benefit from using yeast for bread making, where the carbon dioxide gas is trapped as bubbles and is needed to raise the dough mixture. Anaerobic respiration of yeast is investigated because if yeast respires aerobically, then there would be no pressure change in the tubes as the volume of oxygen consumed will equal the carbon dioxide produced, and no bubbles of carbon dioxide will come through the delivery tube.

In the lab a substance called lime water or aqueous calcium hydroxide ($Ca(OH)_2$)is used to demonstrate the presence of carbon dioxide. When carbon dioxide gas is added to lime water a precipitate (insoluble suspension) of calcium carbonate ($CaCO_3$) is formed which turns the lime water milky. Milky coloured lime water therefore indicates the presence of carbon dioxide. A chemical equation for this is:

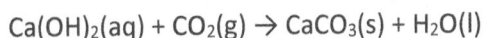

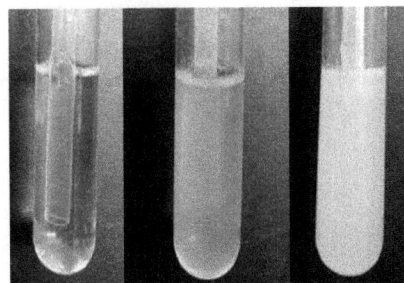

$Ca(OH)_2(aq) + CO_2(g) \rightarrow CaCO_3(s) + H_2O(l)$

NAME:_____CLASS:_____ SKILL: AI/MM

TOPIC: RESPIRATION   SYLLABUS OBJECTIVE: B 3.2

11.   INVESTIGATING THE EFFECT OF TEMPERATURE ON RATE OF ANAEROBIC RESPIRATION OF YEAST

AIM:_____

_____

Why does bread dough rise faster in warmer areas? The rate at which carbon dioxide is produced can be used to measure the rate of anaerobic respiration of yeast. In this experiment you will determine the effect of temperature on the rate of fermentation by investigating gas production by the yeast/glucose mixture with which you have been supplied.

APPARATUS AND MATERIALS:

- Yeast/glucose (5%) mixture
- 1 Beaker (500cm^3)
- 1 Boiling tube
- 1 Syringe (2cm^3)
- Glass rod/ stiirrer
- Plasticine/ anchor
- Water bath (ability to adjust)
- Thermometer
- Ice
- Stopwatch/ clock

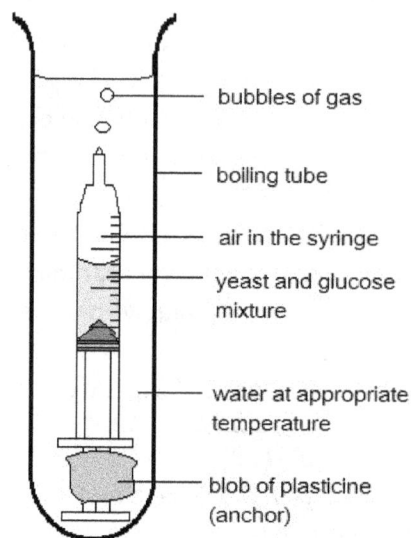

bubbles of gas

boiling tube

air in the syringe

yeast and glucose mixture

water at appropriate temperature

blob of plasticine (anchor)

DRAWING SHOWING SET-UP OF APPARATUS

INSTRUCTIONS:

1. Set up a water bath with temperature at 35°C.
2. Draw in 1cm^3 of yeast/glucose mixture into the syringe followed by 1cm^3 of air.
3. Two-thirds fill a boiling tube with water at 35°C from the beaker and place it in a rack.
4. Attach a blob of plasticine to the plunger of the syringe and drop the syringe into the boiling tube of water. The syringe should sink to the bottom of the tube (see diagram). If this does not happen, remove the syringe and add more plasticine to the plunger.
5. Ensure bubbles of gas are leaving the syringe. Then, after 1 minute, begin to count the number of bubbles evolved in each minute for a total of 5 minutes.
6. Find the average number of bubbles per minute and record on your table.
7. Empty your syringe and rinse it.
8. Stir the yeast and glucose mixture before using again.
9. Repeat the entire experiment with ice water (15°C), room temperature water (25°C) and hot water 45°C and 55°C.

RESULTS:

TABLE SHOWING NUMBER OF BUBBLES PRODUCED BY YEAST CELLS AT DIFFERENT TEMPERATURES.

Temperature (°C)	Number of Bubbles					
	1st minute	2nd minute	3rd minute	4th minute	5th minute	Average
15						
25						
35						
45						
55						

Note: Rate of anaerobic respiration is the average number of <u>bubbles/minute</u>

Plot a graph showing the effect of **temperature** on the **rate of anaerobic respiration.** **Use the checklist:**
  □ *Title*   □ *Scale*   □ *Labels on both axes*   □ *Units in each label*   □ *Plots correct*   □ *Line*

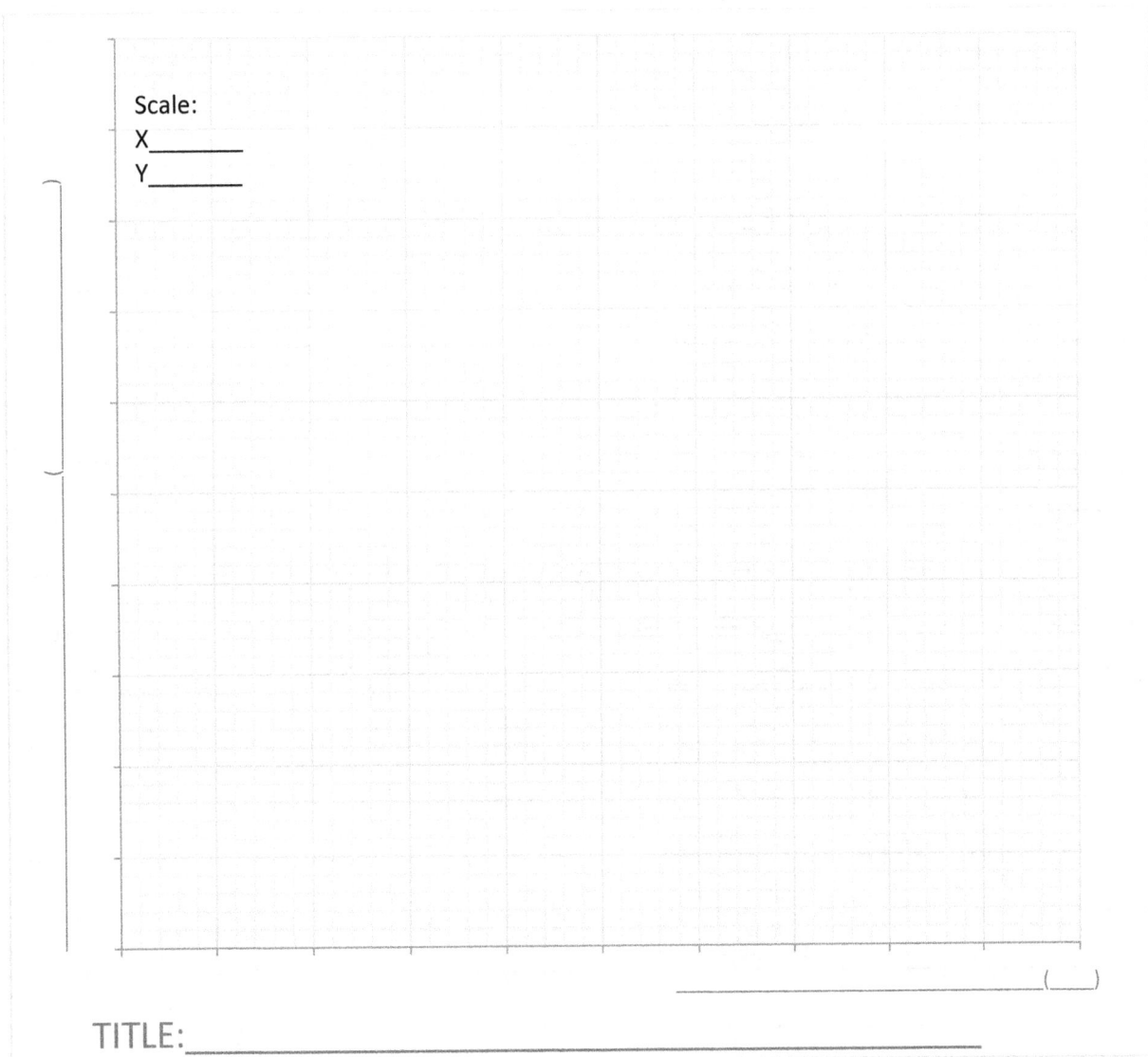

Scale:
  X_____
  Y_____

_____ (____)

TITLE:_____

DISCUSSION

LIMITATIONS _____

_____

_____

PRECAUTIONS _____

_____

_____

SOURCES OF ERROR _____

_____

_____

_____

CONCLUSION _(What was the effect of anaerobic respiration on yeast? How was this identified?)_

_____

_____

_____

REFLECTION _(What was most significant to you? What do you understand about the effect of temperature on a reaction? What is the relevance to your everyday life and using yeast?)_

_____

_____

_____

_____

SAMPLE MARK SCHEME – MM CRITERIA		Mks	SAMPLE MARK SCHEME – AI SPECIFIC CRITERIA		Mks
THERMOMETER	• Bulb immersed in liquid	1	**Background**	Word equation correct and importance	1
	• Stir liquid to distribute heat	1		Products of aerobic vs anaerobic	1
	• Reading taken bulb in liquid	1	**Explanation**	Trend of temperature on rate	1
	• Reading taken at eye level	1		Stated expectation, related to observed results	1
	• Accurate interpretation of scale	1		Quoting values from graph	1
	• Careful handling and storage	1		Improvements to method	1
SYRINGE	• Clean/wash syringe	1	**Limitations, Precautions, Errors**	Correctly explaining 2 limitations or precautions	1
	• Accurate measurement of mixture	1		Correctly explaining 2 S.O.E.	1
	• Attaching plasticine.	1	**Conclusion**	Statement	2
	• Allow time for equilibration before counting bubbles	1		Related to the aim/results	
	TOTAL	10	**TOTAL**		10

DISCUSSION QUESTIONS

1. Using a word equation for anaerobic respiration, identify its importance.
2. What products do yeast cells form when undergoing **aerobic** respiration? What else is formed when yeast cells respire **anaerobically?**
3. In general, why does temperature cause an increase in the rate of reactions? Explain if these were the expected results.
4. Based on the results obtained (observed results), use the graph to identify and explain:
   a. The effect of low temperature (quote) on the rate of CO_2 produced (quote values).
   b. The optimum temperature and the corresponding maximum rate of reaction.
   c. Why rate of reaction decreases beyond a certain temperature. Quote values.
5. If the observed results/ shape of the graph varied from what was expected, explain possible sources of error.
6. How may this experiment be improved to ensure accurate/representative results?
7. Were there any limitations or difficulties in controlling certain aspects of the method? Explain.
8. What precautions were taken to ensure accurate results were obtained? (Mixing and renewing yeast/glucose mixture; maintaining temperatures with water bath; allowing equilibration time)

NOTES – ANAEROBIC RESPIRATION IN YEAST CELLS

Yeast is an organism that has metabolic reactions occurring all the time. Respiration in yeast provides energy to grow and reproduce (forming new cells by budding). The production of bubbles of carbon dioxide gas indicates the rate of respiration occurring.

Respiration is an enzyme controlled reaction with a typical graph on the right. Therefore as temperature increases, respiration will increase, up to 35°C. Increasing temperature causes enzyme and substrate molecules to collide more frequently as they gain kinetic energy. Temperatures beyond this optimum of 35°C can damage and denature enzymes as they are moving too fast, causing them to function less effectively, thereby slowing down respiration. Essentially the enzymes unravel and lose the specific shape of their active site (point where substrates bind to be broken down to products). At temperatures above 50°C, the yeast begin to die due to heat damage so the rate of respiration declines.

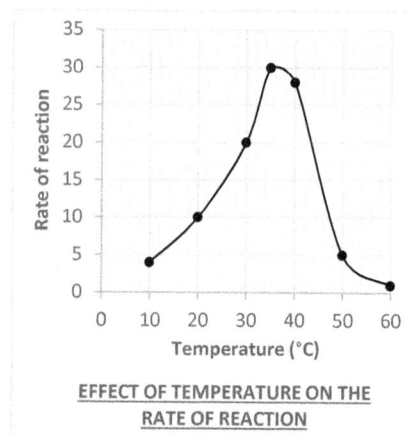

EFFECT OF TEMPERATURE ON THE
RATE OF REACTION

Temperatures below 20°C are too low for reactions to occur and the yeast does not ferment, respiration is also low. Knowledge of these effects of temperature on yeast fermentation is important in allowing efficient fermentation of carbohydrates to leaven (raise) bread and forming alcoholic beverages.

Factors that may affect the ability of yeast to undergo anaerobic respiration include:
- Inactive yeast cells, old yeast cells that could not be activated, will not undergo respiration.
- Incubation period of the yeast with glucose solution. Incubation allows all the oxygen in the test tube to be completely consumed. This is because if yeast respires aerobically, then there would be no pressure change in the tubes as the volume of oxygen consumed will equal the carbon dioxide produced, and no bubbles of carbon dioxide will come through the syringe.

NAME:_____**CLASS:**_____ **SKILL: DR____**

TOPIC: RESPIRATION   SYLLABUS OBJECTIVE: B 3.4

12.   INVESTIGATING GASEOUS EXCHANGE FEATURES OF FISH GILLS

AIM:_____

_____

Do your lungs have anything in common with a fish's gills? Have you ever seen fish gills? They show the characteristic features of gaseous exchange surfaces and their adaptations for efficient gas exchange (thin, large surface area, rich blood supply). Fish gills will be examined and drawn in this lab.

APPARATUS AND MATERIALS:

- Fish gills - one gill bar separated from the 4
- Labelling and annotation guide
- Petri dish with water

- Magnifying glass/ Hand lens
- White tile
- Access to tap – running water

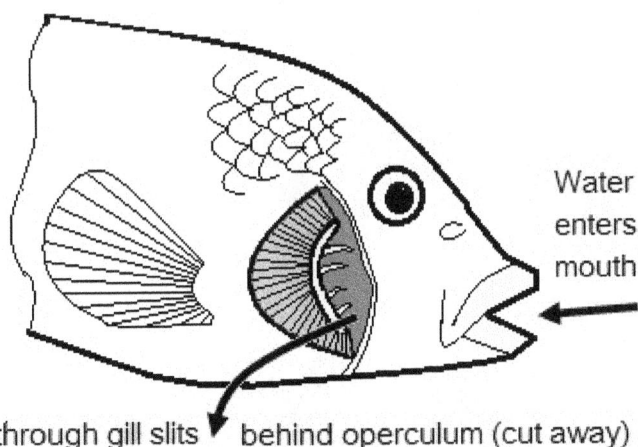

Water enters mouth

Water exits through gill slits   behind operculum (cut away)

DIAGRAM SHOWING LOCATION OF FISH GILLS IN A FISH HEAD (OPERCULUM REMOVED) LATERAL VIEW.

INSTRUCTIONS:

1. Obtain a fresh specimen of a fish gill and immerse in a Petri dish of water on a white tile.
2. Examine, using a hand lens, the external features such as the location and arrangement of the bony gill bar, gill filament and gill rakers.
3. Make a large, labelled drawing of the gill. (LABELS – gill bar, gill filament, gill rakers)
4. Using the guide, add at least 3 annotations (descriptions and functions of labelled parts).
5. Calculate the magnification of your drawing and state it in the title of the drawing. Use the formula:

Magnification of drawing = size of drawing/ real size of specimen

METHOD: (Rewrite your method in past tense in the space below.)

_____

_____

_____

_____

_____

RESULTS: (*Make a large, labelled, annotated drawing of the Fish Gill*)

DRAWING CHECKLIST
CLARITY:
large
clean/ smooth
no shading
ACCURACY
specimen
proportion
LABELLING
parallel
accurate low case
justified
annotation
magnificati
title & view
TOTAL

CALCULATION OF MAGNIFICATION:

$$\text{magnification of drawing} = \frac{\text{size of drawing}}{\text{size of specimen}}$$

$$= \underline{\hspace{2cm}}/\underline{\hspace{1.5cm}}$$

$$= X\,\underline{\hspace{3cm}}$$

TITLE: _____

REFLECTION – *Having observed fish gills, I now (verb)....... /How does this lab relate to your life?*

_____

_____

NOTES: GASEOUS EXCHANGE SURFACES OF AQUATIC ANIMALS – FISH

Gas exchange provides a living organism with a **supply of oxygen gas** for aerobic respiration while, **removing carbon dioxide gas**. The source of this oxygen can be air (in land/ terrestrial animals) or water, as in the case of aquatic organisms such as fish.

Gases move by diffusion. **Diffusion is greater** when 1) the **surface area is large**, 2) the **distance travelled is small** and 3) the **concentration gradient is high**. In general **gas exchange is faster over moist surfaces** that dissolve the gas before diffusing across a membrane. The circulatory system operates in tandem (at the same time) with the gas exchange mechanisms to maintain the concentration gradient.

Fish have an internal gas exchange system that is <u>already moist</u> because fish live in the water. **The fish gills,** inside the mouth of a fish, **are the site of gas exchange**. Gills are covered by an operculum (flap). Ventilation (like breathing) occurs when the fish opens, then closes its mouth. Water with dissolved oxygen flows into the mouth, over the gills and out under the operculum.

gill lamellae

2 rows of gill filaments

Fish gills - structures and <u>adaptations for efficient gas exchange</u>:

- Each gill is made up of **four** bony/cartilaginous **gill arches**. This increases surface area.
- Each gill arch/ gill bar has **2 rows of gill filaments** made up of hundreds of **flaps** that are **thin and flat.** Numerous gill filaments increases surface area.
- Gill filaments have tiny folds called **lamellae** (increasing surface area again)
- Gill lamellae contain a network of **capillaries**. A rich blood supply ensures a steep diffusion gradient.
- Blood flows through the blood capillaries in the **opposite direction** to the flow of water. This maintains the diffusion gradient and increases gas exchange.

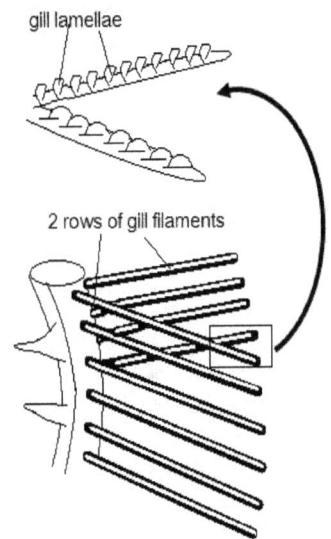

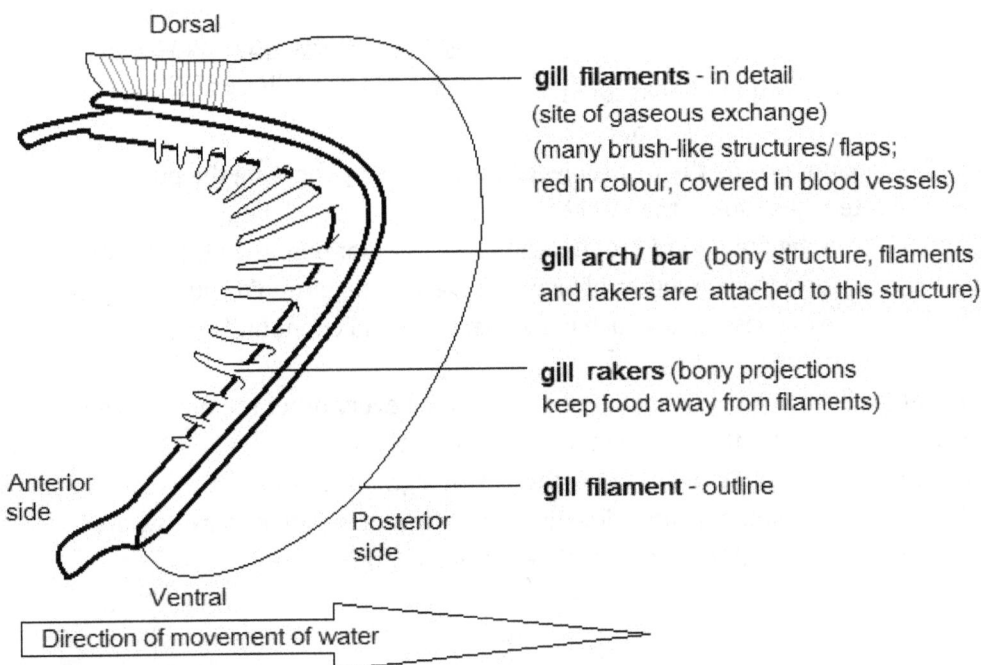

Dorsal

gill filaments - in detail
(site of gaseous exchange)
(many brush-like structures/ flaps;
red in colour, covered in blood vessels)

gill arch/ bar (bony structure, filaments and rakers are attached to this structure)

gill rakers (bony projections keep food away from filaments)

gill filament - outline

Anterior side

Posterior side

Ventral

Direction of movement of water

Interesting fact: Gill rakers indicate the diet of the fish:

- *Predatory fish have short widely spaced rakers, whereas,*
- *Fish that eat smaller prey have longer, thinner and numerous gill rakers.*

<u>DIAGRAM SHOWING A THE EXTERNAL FEATURES OF A TYPICAL FISH GILL AND THEIR FUNCTIONS</u>

NAME:_____CLASS:_____ SKILL:_AI____

TOPIC: RESPIRATION   SYLLABUS OBJECTIVE: B 3.3, 4.2, 4.3

13.   INVESTIGATING THE EFFECT OF EXERCISE ON BREATHING RATE AND PULSE RATE

AIM:_____

_____

Your pulse indicates how fast your heart beats. It can be measured along major arteries in the wrist or neck. Did you know that pulse rate is related to the breathing rate? An increased heart rate can stimulate increased breathing rate because of oxygen demand by cells. Are the changes in these rates the same, given different levels of athletic ability? The purpose of this experiment is to answer some of these questions while investigating the overall effect of exercise on breathing rate and pulse rate.

APPARATUS AND MATERIALS:

- Stopwatch
- 3 students – athletic, moderate exercise, non-athletic – to exercise
- 3 students to assist in taking pulse rate

Place two fingers on the wrist, just below the thumb to feel the radial artery.

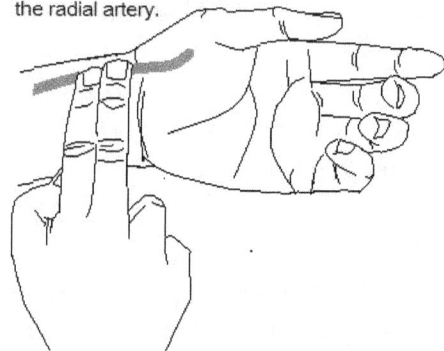

DIAGRAM SHOWING HOW TO FIND THE PULSE IN THE RADIAL ARTERY

METHOD:

1. Select 3 students of varying athletic ability from within the class to be the test subjects.
2. Allow each individual to sit quietly for 2 minutes of quiet breathing.
3. In the third minute, each individual should count his/her own number of breaths (number of chest rises) while another person records the test subjects' pulse rates by checking the pulse at the neck or in the radial artery, just under the thumb.
4. Repeat the count for both the number of breaths and pulse rate per minute in the fifth minute.
5. Have each individual perform three minutes of vigorous exercise (e.g. running, jumping jacks.).
6. Immediately following the exercise, allow each individual to sit and count his/her number of breaths while his/her pulse rate is taken.
7. Continue the count for both pulse rate and number of breaths every other minute until each returns to the resting pulse rate (this may take several minutes).
8. Record both pulse rate and breathing rate in the table. Plot a graph to compare the pulse rates of the students in beats per minute against time. Identify with arrows on the x axis where exercise started and stopped, and the recovery times for each individual

RESULTS:

TABLE SHOWING_____

Time/ min	Athletic individual		Individual with moderate exercise levels		Non athletic individual	
	Number of breaths per minute	Pulse rate/ beats per minute	Number of breaths per minute	Pulse rate/beats per minute	Number of breaths per minute	Pulse rate/beats per minute
3						
5						
Vigorous exercise performed for three minutes – running/ jumping jacks, etc…						
9						
11						
13						
…						

Plot a graph below using the checklist:
  □ *Title*   □ *Scale*   □ *Labels on both axes*   □ *Units in each label*   □ *Plots correct*   □ *Line*

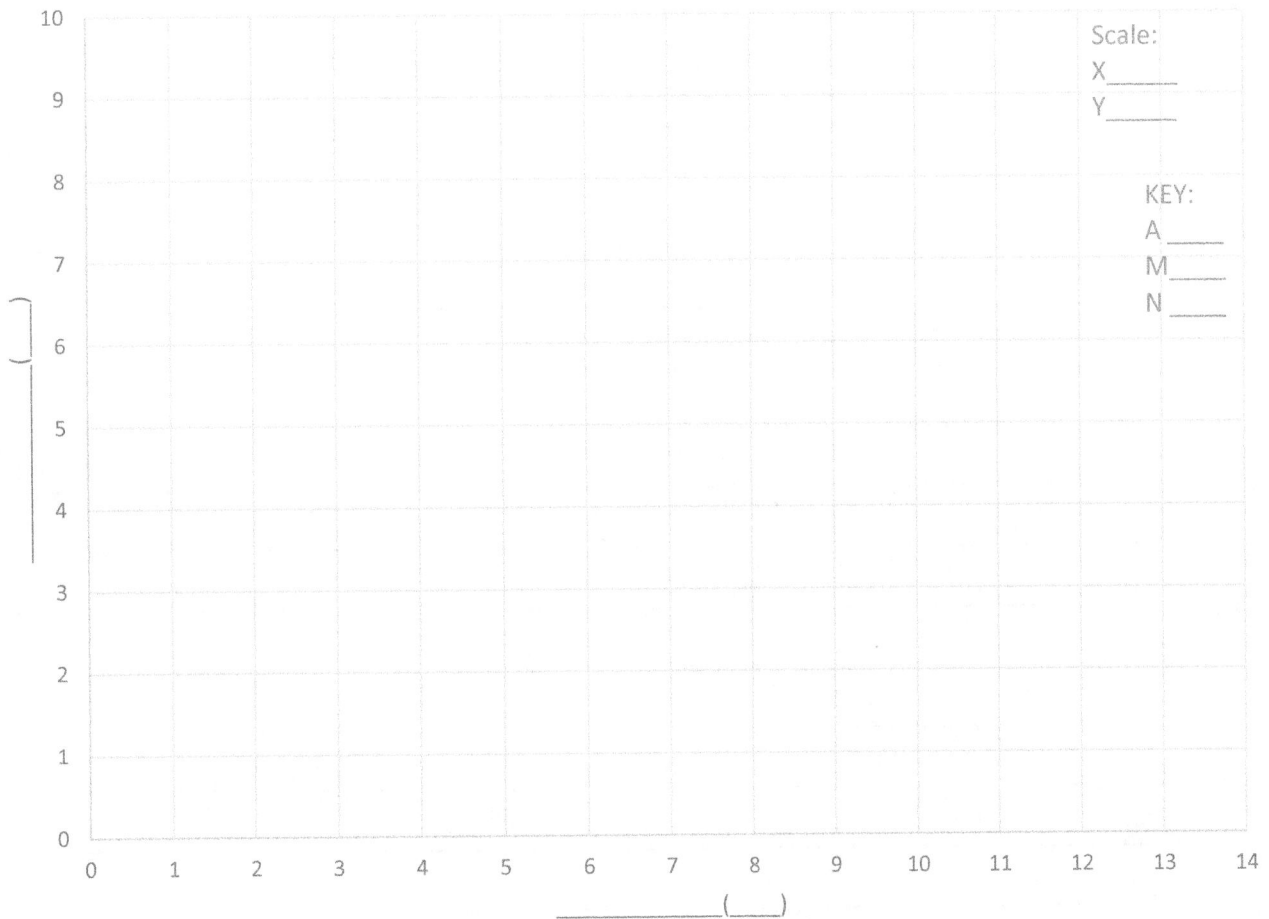

Scale:
X_____
Y_____

KEY:
A _____
M_____
N _____

TITLE: _____

DISCUSSION

LIMITATIONS _____

_____

_____

PRECAUTIONS _____

_____

_____

SOURCES OF ERROR _____

_____

_____

_____

CONCLUSION *(What was the overall effect of exercise on breathing and pulse rate? Did it matter if the person was an athlete/ exercised regularly?)*

_____

_____

_____

_____

_____

REFLECTION *(What was most significant to you? What do you understand about exercise and heart health? How will you take this knowledge and use it in the future?)*

_____

_____

_____

_____

_____

_____

_____

DISCUSSION POINTS

1. Explain the importance of breathing especially during exercise. Link oxygen intake to the removal of carbon dioxide waste and lactic acid from muscles removal.
2. Explain why the pulse rate is expected to also increase with exercise.
3. Using the graph compare the number of breaths per minute for the three individuals by identifying the highest and lowest. Was there any correlation with the exercise?
4. Identify the recovery times for each individual, saying which was the longest, and which the shortest. Explain the most likely reasons for these results.
5. How can the experiment be improved to achieve more accurate results?
6. Limitations: Explain any mistakes made during the experiment or uncontrollable factors that could have affected the results (e.g. student(s) not performing exercise to the same intensity). Identify any variables that were not controlled that could have affected the results (Hint: Individuals may not be the same height/weight/Body Mass Index (BMI), etc., genetic factors/predisposition to cardiovascular disease could affect breathing and pulse rate).

CONCLUSION:

State what effect exercise had on the pulse rate and breathing rate of students. Identify the category of person (athletic, moderate exercise or non-athletic) which had more of an effect.

NOTES - EFFECT OF EXERCISE ON BREATHING AND HEART RATE

Pulse is an indication of the heart rate. Arteries expand every time the ventricles contract and pumps blood out the heart. Exercise increases the need for energy in the muscles; therefore more food and oxygen is needed, so both the pulse rate and breathing rate increases. Furthermore, as more aerobic respiration occurs, carbon dioxide that forms must be removed, thus increasing breathing rate.

Exercise in general causes the pulse rate and number of breaths per minute to increase, especially in a non-athletic individual. A person who is athletic and exercises regularly may observe less of an effect, even though these do increase. The time it takes for pulse and breathing rate to return to normal after exercise has stopped is called the **recovery time**. More athletic persons have shorter recovery times. This is because:

- Exercise increases the heart's capacity and strength
- Individuals who regularly exercise have fewer heartbeats as the heart is more efficient in pumping blood to the necessary muscles that need oxygen
- Exercising daily trains the heart causing decreased resting heart rate as well as the exercise heart rate.

Daily exercise is therefore important in preventing heart diseases such as heart attack, angina (chest pain) and even stroke. Doctors recommend 150 minutes of moderate exercise or 75 minutes of vigorous exercise per week (or a combination of these). A minimum, easy goal should therefore be thirty minutes a day, five times a week. Lowered blood pressure and cholesterol, reduced risk for cardiovascular illness and even weight loss and some building of muscle mass are some of the merits of daily aerobic exercise. Furthermore, exercise reduces stress by releasing feel-good hormones known as endorphins which can also improve heart health.

NAME:_____**CLASS:**_____   **SKILL: ORR/AI**

TOPIC: TRANSPIRATION   SYLLABUS OBJECTIVE: B 4.8, 4.9

14.   OBSERVING THE EFFECT OF WIND ON THE MOVEMENT OF WATER UP A CELERY STEM

AIM:_____

_____

How does the weather affect plants? Should farmers consider the wind when deciding where to plant? The hypotheses "transpiration is faster in moving air" and "wind increases the rate at which water moves up the xylem" are two possible hypotheses to be tested. This investigation relates the movement of coloured water up a celery stem, to the transpiration water loss from the leaf.

APPARATUS AND MATERIALS:

- 2 Conical flasks
- Measuring cylinder
- 2 Beakers
- 2 Scalpels
- Stop clock
- 2 Rulers
- Labels

- Standing fan
- 2 White tiles
- Hand lens/ magnifying glass
- Blue solution
- 2 Celery stems (fresh with leaves)
- Distilled water

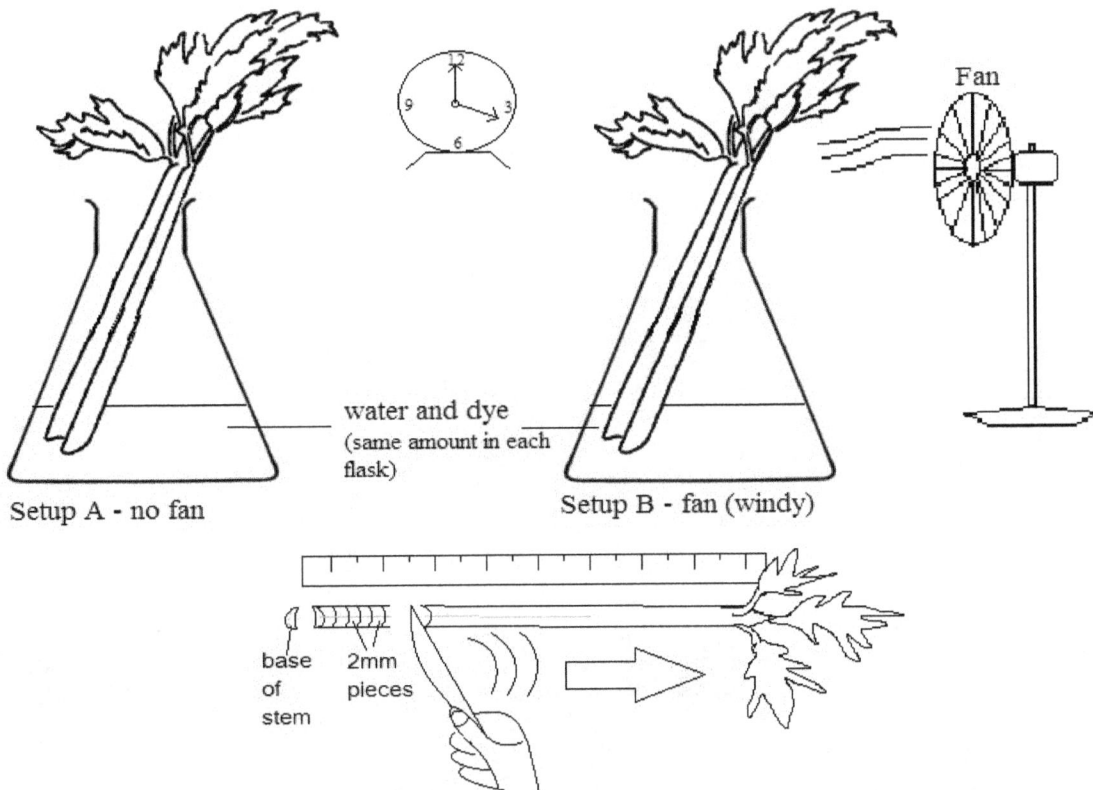

water and dye
(same amount in each flask)

Setup A - no fan

Setup B - fan (windy)

base of stem   2mm pieces

DIAGRAM SHOWING SET-UP A AND B – TO INVESTIGATE THE EFFECT OF WIND ON MOVEMENT OF WATER UP A CELERY STEM

INSTRUCTIONS:

1. Label two conical flasks A and B and place 40ml of blue solution into each.
2. Obtain 2 similar stems of celery, hold each stem under water and cut off the base using a scalpel.
3. Place one stem in flask A in front of a fan at high speed.
4. Place the other stem in flask B on the bench top away from any wind.
5. Leave both flasks for 30 minutes and then measure and record the length of the celery stem from the end to the base of the first set of leaves.
6. Cut the stems in cross section (2mm at a time from the base) and determine how far the blue liquid moved up the stems. Record this value.
7. Calculate the transpiration rate for each celery stem and record in the table.
   Note: Transpiration Rate = distance dye moved per minute.
8. Use a thin cross-section of the celery stem with the blue dye in the xylem vessels and draw what you observe with a magnifying glass.

Rewrite your method in past tense in the space below or on a separate page.

METHOD:

RESULTS:

*Draw a table with TITLE to collect the results based on the **different conditions** each stem receives.*

TITLE:_____

```
┌─────────────────────────────────────────────────────────────┐
│                                                             │
│                                                             │
│                                                             │
│                                                             │
│                                                             │
│                                                             │
│                                                             │
│                                                             │
│                                                             │
│                                                             │
│                                                             │
│                                                             │
│                                                             │
│                                                             │
│                                                             │
└─────────────────────────────────────────────────────────────┘
```

***Draw** a cross-section of the celery stem and **label it** to clearly show the parts that were coloured blue.*

```
┌─────────────────────────────────────────────────────────────┐
│                                                             │
│                                                             │
│                                                             │
│                                                             │
│                                                             │
│                                                             │
│                                                             │
│                                                             │
│                                                             │
│                                                             │
│                                                             │
│                                                             │
│ TITLE:_____ │
└─────────────────────────────────────────────────────────────┘
```

DISCUSSION

LIMITATIONS _____

_____

_____

_____

PRECAUTIONS _____

_____

_____

SOURCES OF ERROR _____

_____

_____

_____

CONCLUSION _____

_____

_____

_____

_____

_____

_____

REFLECTION _____

_____

_____

_____

_____

_____

CSEC SAMPLE MARK SCHEME - ORR CRITERIA				MARK	T. Mk.
O – **Observations**	s o c/d	• Significant changes noted (1) • Original and final conditions compared (1) • Control noted OR diagram (1)		**3**	
RTG – **Recording**	t u e	**TABLE** • Title – above, in capitals, underlined (1) • Column & row headings(with units) (1) • Enclosed and neat (1)	t u p **GRAPH** • Title below, capitals, underlined (1) • Both axes labelled with units (1) • Accurate plots (1)	**3**	
R – Reporting	a r g s	• Aim in capital letters (1) • Reflection appropriate (1) • Acceptable language and expression – subject-verb agreement/ grammar (1) • Spelling correct throughout with 0-3 errors (1) *Note: > 3 grammatical errors = 0 marks. AND > 3 spelling errors = 0 marks*		**4**	
		TOTAL		**10**	

CSEC SAMPLE MARK SCHEME - AI GENERAL CRITERIA			SPECIFIC CRITERIA – Transpiration experiment	MARK	T. Mk.
B- Background	d s	• Define key related terms (1) • Statement of relevant theory (1)	Transpiration – evaporation of water through stomata in leaves How xylem works – 1 process (capillarity/ adhesion and cohesion)	2	
E -Explanation	t c u m	- Trends and patterns identified (1) - Compare actual results with expected results (1) - Use data to support explanations (1) - Modification/ improvement to existing method (1)	Higher transpiration identified Expectation that wind (not fan) causes higher transpiration due to loss of moisture just outside leaf Quote results values of transpiration rate or distance dye moved up stems Improvement explained	4	
LSP -Limitations Sources of error/ Precautions		- Identify at least 2 limitations with explanations - Identify at least 2 precautions/ sources of error with explanations	Thicker stems have more xylem vessels; unhealthy leaves closed stomata Same length and size of stems/ healthy or undamaged stems, etc.	2	
C- Conclusion	s r	- Statement - Related to aim	Water moved up the stem higher....Due to wind.	2	
		TOTAL		**10**	

DISCUSSION QUESTIONS:

1. What is transpiration and how does it take place? How is it useful for plants?
2. Name the tissue involved in water transport and state one adaptation for this purpose.
3. Explain one way water moves up the xylem – cohesion, adhesion, capillarity or transpiration pull.

4. Explain with reasons which set-up is expected to have the higher transpiration rate.
5. Compare your expectation with the observed results, saying which celery stem showed the higher transpiration rate (**use the recorded results** - quote values from the table).
6. Identify any sources of error which may have made the results differ from the expected results.
7. This experiment assumes the rate of movement of dye is equal to the rate of transpiration, but is this always true? What else does a plant use water for?

8. How can the experiment/ method be improved for more accurate results?
9. Identify and explain some precautions in carrying out the experiment; explain any sources of error.

NOTES - TRANSPIRATION

Transpiration is the **loss of water by evaporation** via the leaf stomata. Transpiration inadvertently occurs when stomata are open This is needed for gas exchange of carbon dioxide for photosynthesis. However, transpiration is **important** for the following reasons:

- It pulls (suctions) water up to the leaves via the xylem tissue.
- The moving water carries dissolved mineral salts.
- The evaporation of water cools the plant.

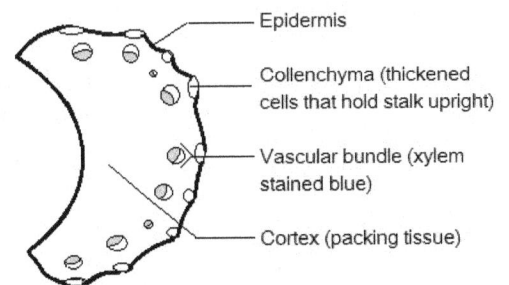

Epidermis

Collenchyma (thickened cells that hold stalk upright)

Vascular bundle (xylem stained blue)

Cortex (packing tissue)

INTERNAL STRUCTURES OF A CELERY STEM IN C.S.

Transpiration is affected by environmental factors as well as leaf morphology factors.

Environmental factors include:
- Relative humidity
- Air movement/ wind
- Temperature
- Water supply
- Light intensity

Factors of the leaf morphology:
- Leaf surface area
- Thickness of epidermis and cuticle
- Stomatal frequency (# stomata per surface area)
- Stomatal size
- Stomatal position (lower and/ or upper epidermis)

Increased air movement as created by the fan blowing the leaves, increases the rate of transpiration. This is because the boundary layer of air that is just outside the leaf (with water vapour) is removed. A steeper diffusion gradient for water vapour causes water to diffuse out of the leaf. Too much water loss from transpiration causes wilting of plants (when all the cells in the leaf/ plant become flaccid).

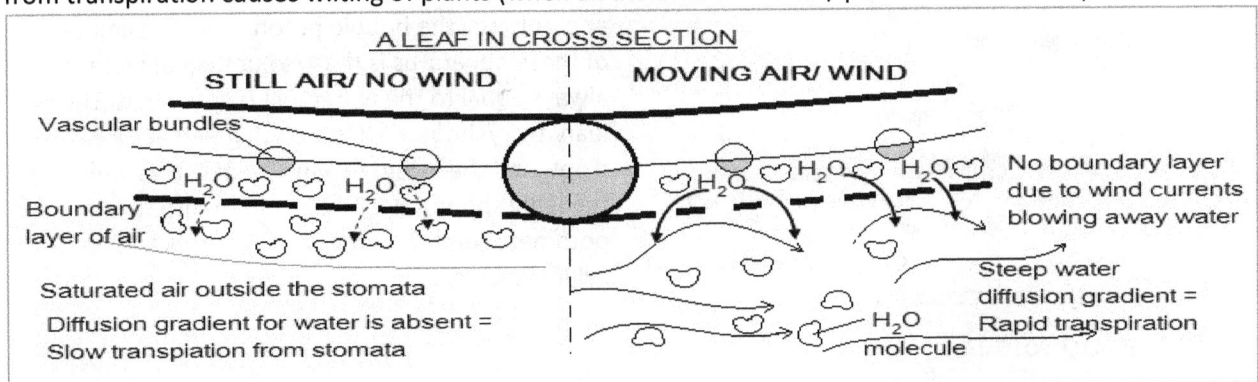

A LEAF IN CROSS SECTION

STILL AIR/ NO WIND   MOVING AIR/ WIND

Vascular bundles

Boundary layer of air

No boundary layer due to wind currents blowing away water

Steep water diffusion gradient = Rapid transpiration

Saturated air outside the stomata

Diffusion gradient for water is absent = Slow transpiation from stomata

H_2O molecule

- It is **assumed** that the rate at which a plant takes up water depends on the rate at which it is lost from the plant - the transpiration rate. However, measuring water uptake is an indirect measurement of transpiration. This is because water can be used to keep cells turgid or for photosynthesis, as such, not all water taken up the stem is lost via transpiration.

- How can this transpiration experiment be **improved**?
  - *Leave the stems for a longer time period and then measure.*
  - *Add a paper collar/ oil around the stem to reduce evaporation of blue water from the flask.*
  - *Repeat the set-up several times.*
    - *Measure the initial volume of dye solution and at fixed intervals.*
    - *Try using an entire plant (with a translucent stem – shining bush) with intact roots.*

  - *What were some **precautions** for a successful transpiration experiment?*
    - *Ensure stem lengths are the same at the start.*
    - *Ensure celery stem and leaves are healthy (not already wilted or water stressed.) Wilted leaves have closed stomata so no water will be lost.*
    - *Ensure that there was no puncture or bruise on the stem, since damaged xylem vessels prevent capillarity action.*

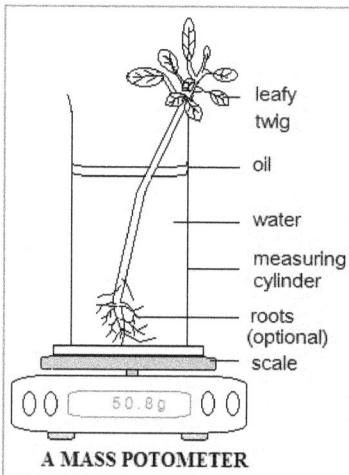

A MASS POTOMETER

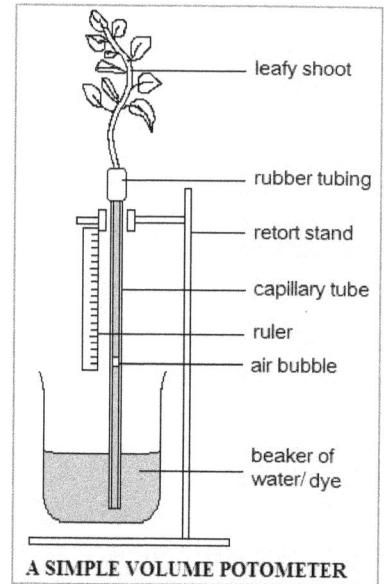

A SIMPLE VOLUME POTOMETER

OTHER APPARATUS TO MEASURE TRANSPIRATION.

Transpiration can also **be measured using a simple potometer**. There are two types

1) **A weight/ mass potometer** – <u>measures the rate of loss of mass from a plant</u> or leafy twig over several hours. Most of the mass lost by the plant will be due to the water evaporating from the leaves during transpiration.

TIME	Mass of apparatus (g)	
	With Fan	Without Fan
At the start		
After 1 hour		
Loss in mass		

In general $1cm^3$ of water weighs 1g.

2) **A volume potometer** - <u>finds the rate of uptake of water</u> by a leafy shoot using a capillary tube.

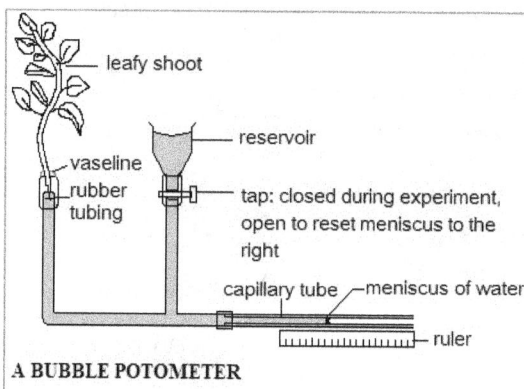

A BUBBLE POTOMETER

These could be simple tubes or more complex apparatus as the bubble potometer. The limitations of these apparatus is that the uptake of water is not always equal to the water lost by the plant. The leafy twig/shoot itself is a cutting and not a whole plant, and the xylem may be damaged as well. Note: the leafy shoot must be cut and fit into the potometer **under water**, so no air gets trapped in xylem vessels.

NAME:_____ **CLASS:**_____ **SKILL: DR_____**

TOPIC: TRANSPORT IN PLANTS/ STORAGE   SYLLABUS OBJECTIVE: B 4.13, 4.11

15.   INVESTIGATING PLANT STORAGE ORGANS AND THEIR BUDS

AIM:_____

_____

Plants store the food they make, some of which we can eat. Food can be stored in underground stems, roots, bulbs and even above ground stems as in the sugarcane. These organs are the sinks for food translocated from leaves and they overcome the need for continuous food supply. These storage organs can be used to form new plants from the buds by asexual reproduction. In this experiment you will become familiar with the parts of the plant that form storage organs such as the potato stem tuber, carrot tap root and onion bulb.

APPARATUS AND MATERIALS:
- Potato tuber - cleaned
- Carrot
- Onion
- Magnifying glass
- Knife
- White tile
- Labelling and annotation guide
- Mechanical pencil/ sharpened pencil
- Ruler
- Clean eraser
- Calculator

INSTRUCTIONS:
1. Obtain the plant storage organs – potato, carrot and onion bulb.
2. Cut the onion longitudinally with a knife through the centre.
3. Examine the internal features of the onion bulb as seen in section.
4. Examine the external features of the potato and carrot. Be sure to note the growing points or buds and any leaf scars.
5. Make large, labelled drawings of each specimen.
6. Using the guide, add **at least 3** annotations (description and functions of labelled parts).
7. Calculate the magnification of your drawing and state it in the title of the drawing. Use the formula:

Magnification of drawing = size of drawing/ real size of specimen

METHOD: *(Rewrite your method in the past tense in the space below.)*

_____
_____
_____
_____
_____
_____
_____
_____
_____

RESULTS: (*Make a large, labelled, annotated drawing of the potato stem tuber*)

DRAWING CHECKLIST
CLARITY:
large
clean/ smooth
no shading
ACCURACY
specimen
proportion
LABELLING
parallel
accurate lower case
justified
annotations
magnification
title &view
TOTAL

CALCULATION OF MAGNIFICATION:

$$\text{magnification of drawing} = \frac{\text{size of drawing}}{\text{size of specimen}}$$

$$= \underline{\hspace{2cm}} / \underline{\hspace{2cm}}$$

$$= \underline{\text{X}\hspace{3cm}}$$

TITLE: _____

(Make a large, labelled, annotated drawing of the carrot tap root)

DRAWING CHECKLIST
CLARITY:
large
clean/ smooth
no shading
ACCURACY
specimen
proportion
LABELLING
parallel
accurate lower case
justified
annotation
magnificatn
title & view
TOTAL

CALCULATION OF MAGNIFICATION:

$$\frac{magnification}{of\ drawing} = \frac{size\ of\ drawing}{size\ of\ specimen}$$

$$= \underline{\hspace{2cm}} / \underline{\hspace{2cm}}$$

$$= \underline{X \hspace{3cm}}$$

TITLE: _____

(Make a large, labelled, annotated drawing of the onion bulb cut longitudinally)

DRAWING CHECKLIST	
CLARITY:	
large	
clean/ smooth	
no shading	
ACCURACY	
specimen	
proportion	
LABELLING	
parallel	
accurate lower case	
justified	
annotation	
magnificati	
title & view	
TOTAL	

CALCULATION OF MAGNIFICATION:

$$magnification \text{ of drawing} = \frac{size \text{ of drawing}}{size \text{ of specimen}}$$

$$= \underline{\hspace{2cm}} / \underline{\hspace{2cm}}$$

$$= X \underline{\hspace{3cm}}$$

TITLE: _____

REFLECTION – *This lab gave you a better understanding or appreciation for...?*

_____

_____

NOTES: DESCRIBING PLANT STORAGE ORGANS/ FOOD STORAGE IN PLANTS

Food that is made in the leaves of the plants is translocated (moved) via the phloem tissues to sinks. Sinks may be flowers, fruits and roots. These parts metabolise the food for growth or form storage organs usually underground. These underground structures may be roots as in carrot, cassava and sweet potato; modified stems as in the Irish potato and even modified storage leaves as in the onion bulb. The importance of storage organs can be:

i. Plants overcome the need for continuous food manufacture.
ii. Plants may develop/ stay alive during times of scarcity or drought.
iii. Asexual reproduction/ vegetative propagation of new plants.

POTATO TUBER:

- Potato plants have an underground stem.
- The underground stems become swollen with stored starch and are called tubers.
- There is a scar on the tuber which was the point where it was attached to the parent plant.
- The little "eyes" that are on the potato are actually tiny lateral buds (the eye) with scale leaf scars (the eyebrow).
- Each bud has the potential to grow and become a new plant.
- When these new plants form, the existing potato becomes soft and wrinkled as food is used up to form these new stems, roots and leaves.

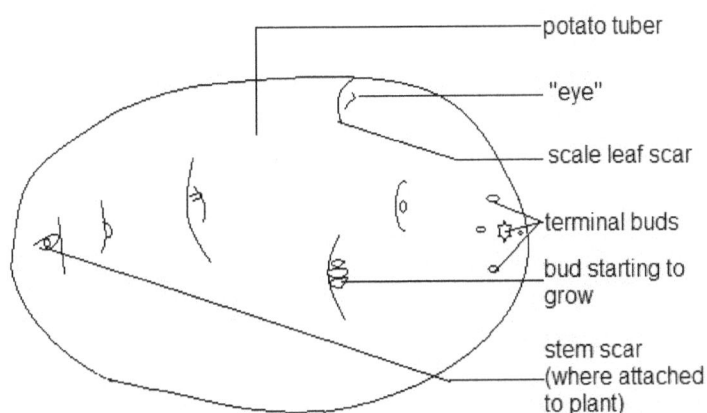

THE EXTERNAL FEATURES OF A POTATO TUBER
Solanum tuberosum

ONION BULB

- Bulbs have very short stems.
- Closely packed storage leaves circle the stem.
- New leaves form in the centre of the bulb.
- The inner fleshy leaves store sucrose.
- The oldest leaf dries up, become brown, thin and papery to protect inner leaves. This is called a tunic.
- The adventitious roots are dried and shrivelled.
- The lateral (axillary) buds will produce new plants called bulbils.
- Bulbs are considered vegetative structures as they can form new plants from a parent plant.

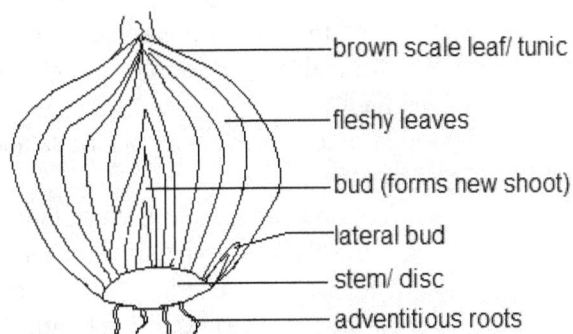

ONION BULB - LONGITUDINAL SECTION
Allium cepa

CARROT TAP ROOT

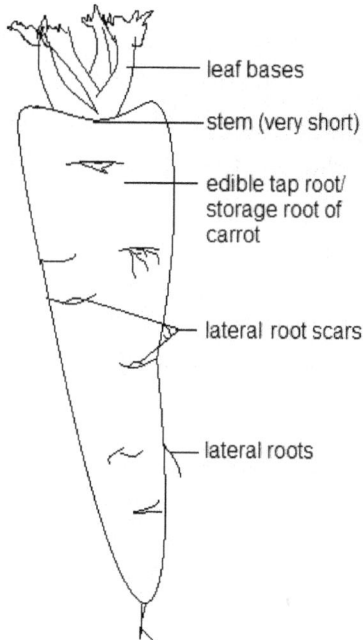

THE EXTERNAL FEATURES OF
A CARROT TAP ROOT
Daucus carota sativus

- Carrots are thickened tap roots with very few lateral roots.
- They are usually orange and conical/ cylindrical.
- They have a very short inconspicuous stem at the top that produces a cluster of leaves.
- The periderm or skin is composed of suberin and other waxy substances.
- They grow for only 2 years – biennials.
- In the first year they make food in the leaves and store the sugars in their tap roots as starch.
- In the second year, they use the stored food to produce flowers (called bolting) which form seeds – used to plant more carrots.
- Carrots are harvested before they bolt, that is, after the first year of growth.

SUGAR CANE

- Sugar cane is a perennial member of the grass family.
- It is tall and looks like a bamboo cane.
- It grows mainly in the tropics.
- Leaves at the top of the stem make sugars which is stored in the stems.
- It is widely cultivated to provide white sugar and brown sugar.
- Molasses, a by-product of the refining process, is used to make syrups and can be fermented and distilled to produce rum and ethanol (a motor fuel)
- Each joint is made up of a node and an internode.
- The node is where the leaf attaches to the stalk, the leaf scars remaining on the storage stem.
- New plants can be formed by cuttings that include a node in the root band.

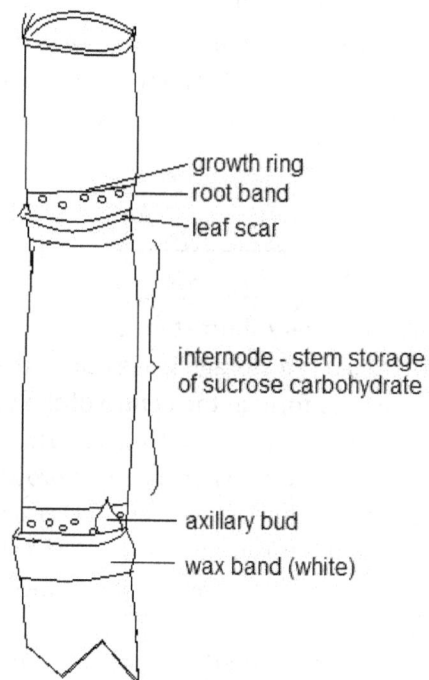

THE EXTERNAL FEATURES OF A
SUGAR CANE STEM
Saccharum officinarum

NAME:_____**CLASS:**_____ **SKILL: MM/ORR**

TOPIC: TRANSPORT IN PLANTS/ STORAGE   SYLLABUS OBJECTIVE: B 4.12, 4.13

16.   IDENTIFICATION OF THE BIOMOLECULES IN STORAGE ORGANS – POTATO & RED BEANS

AIM:_____

_____

Which food is more nutritious for a vegetarian, potato, which is a root, or red beans which are seeds? Plants store food in different parts – roots, stems, leaves and seeds, and humans can eat these for nutrition. This experiment uses the simple food tests to examine specifically what is stored in storage organs.

APPARATUS AND MATERIALS:

- Mortar and pestle
- 9 Test tubes
- 2 Small beakers
- Test tube holder
- 3 Measuring cylinders
- Syringe 2cm^3
- Large beaker
- Bunsen burner

- Gauze and tripod stand
- Labels
- Stirring rod
- Spatula
- Stop clock
- Forceps

- Distilled water
- Benedict's solution
- Sodium hydroxide (NaOH)
- Copper sulphate (CuSO$_4$)
- Iodine solution
- Seeds – red bean
- Potato

DIAGRAM:

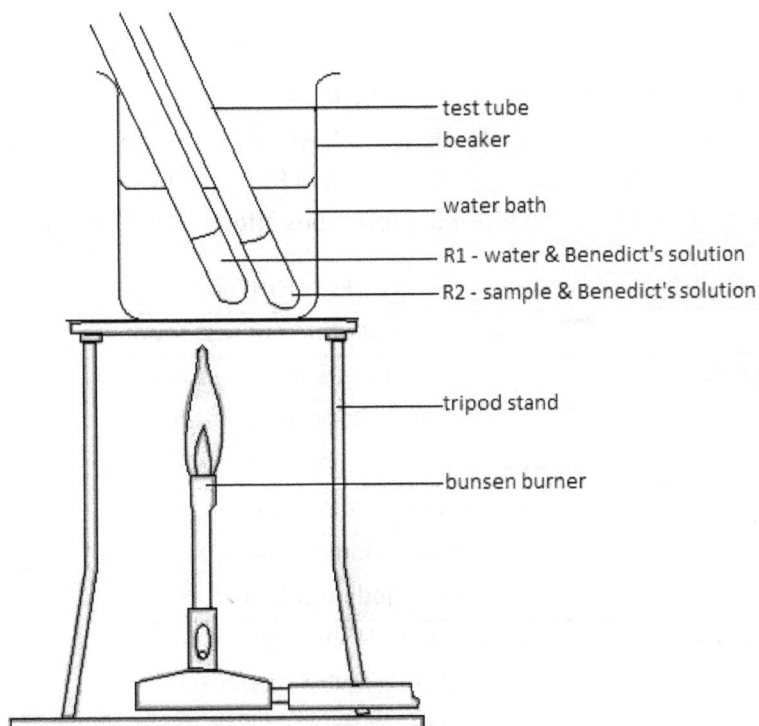

DIAGRAM SHOWING THE SET UP OF WATER BATH FOR REDUCING SUGAR TESTS

METHOD:
1. Label two beakers Potato and Seed respectively.
2. Prepare the potato as follows:
   a. Place small cubes of peeled potato in a blender with enough distilled water to cover.
   b. Blend to a puree or suspension and place in the beaker labelled Potato.
3. Prepare the seeds as follows:
   a. Remove the testas from soaked peas with a forceps.
   b. Place in a mortar and crush with a pestle until a paste form.
   c. Add this paste to the beaker labelled Seed along with $10cm^3$ of distilled water.
   d. Stir this mixture with a clean glass rod and set aside.
4. Label 9 test tubes – R1, R2, R3; S1, S2, S3 and P1, P2, P3
5. Add $2\ cm^3$ distilled water to each test tube labelled R1, S1 and P1 respectively.
6. Add $2\ cm^3$ of potato mixture to R2, S2 and P2 respectively.
7. Add $2\ cm^3$ of seed mixture to R3, S3 and P3 respectively.

(A) BENEDICT'S TEST (REDUCING SUGAR TEST):
8. Add $2\ cm^3$ of Benedict's solution to each of the $2\ cm^3$ of test sample (R1, R2 and R3).
9. Record the initial colours, then place all test tubes in a boiling water bath for 2 minutes.
10. Observe and record the changes in the colours during the 2 minutes and the final colours.

(B) STARCH TEST (IODINE TEST)
11. Record the initial colours of S1, S2 and S3.
12. Add 3 – 5 drops of iodine solution to each test tube S1, S2 and S3 respectively and observe and record the finial colours of each test tube.

(C) PROTEIN/ BIURET TEST
13. Record the initial colours of P1, P2 and P3 in the table.
14. Add $1\ cm^3$ of dilute sodium hydroxide (NaOH) followed by 4 drops of 1% Copper Sulphate solution ($CuSO_4$) to each test tube sample P1, P2 and P3 respectively.
15. Observe and record the final colours of the test tubes after about 1 minute.

TABLE SHOWING CONTENTS OF EACH TEST TUBE AND TEST DONE.

Name of Test	Sample and Contents of test tube
Reducing sugar test	R1 – Water + Benedict's solution
	R2 – Potato + Benedict's solution
	R3 – Seed + Benedict's solution
Starch/ Iodine test	S1 – Water + Iodine solution
	S2 – Potato + Iodine solution
	S3 – Seed + Iodine solution
Protein/ Biuret test	P1 – Water + NaOH + 1%$CuSO_4$
	P2 – Potato + NaOH + 1%$CuSO_4$
	P3 – Seed + NaOH + 1%$CuSO_4$

RESULTS:

In the space provided, construct a table to record your observations (initial and final colours of solutions and inferences – positive or negative for the biomolecule).

TITLE:_____

DISCUSSION

_____

_____

_____

_____

_____

_____

_____

_____

_____

_____

_____

_____

_____

_____

_____

_____

_____

_____

LIMITATIONS

_____

_____

PRECAUTIONS

_____

SOURCES OF ERROR

_____

CONCLUSION – *What was found out in this lab? Potato stores...... whereas beans store......* _____

_____

REFLECTION – *What impact did this experiment have on you and your everyday life?*

_____

_____

_____

DISCUSSION QUESTIONS:

1. What are plant storage organs? (Hint – discuss roots, stems, leaves and seeds)
2. Give 3 reasons why nutrients are stored in storage organs.

3. Identify the nutrient(s) expected to be found stored in (a) potato and (b) seeds.
4. Based on the results,
   o What was stored in a potato and why?
   o What was stored in seeds and why?
5. Are there any other nutrients that may be present in the storage organs but were not tested? (Hint- lipids in the seeds).
6. Why was the potato blended and the seeds crushed before conducting the food tests?

7. What precautions were taken in the experiment? (Hint – washing the measuring cylinders between each test conducted/ placing the test tubes into the water bath at the same time.)
8. Identify any sources of error in carrying out the experiment.
9. What were possible limitations for getting accurate results?

Criteria (MM) - Using the measuring cylinder + reagent bottles		Maximum Marks	Teacher Marks
MEASURING CYLINDER	• Cylinder on flat surface	1	
	• Readings taken at eye level	1	
	• Readings taken at the bottom of the meniscus	1	
	• All liquid poured out completely from cylinder	1	
	• Cylinder is cleaned and washed between uses • and at end of lab	2	
REAGENT BOTTLES	• Select the correct reagent – read the label	1	
	• Pour away from the label	1	
	• Care of stopper – upside down	1	
	• Equal size drops released from the dropper	1	
TOTAL		**10**	

SAMPLE MARK SCHEME - ORR GENERAL CRITERIA				Max. Mark	T. Mk.
O – OBSERVATIONS	s o c/d	• Significant changes noted (1) • Original and final conditions compared (1) • Control noted OR diagram (1)		3	
RTG – RECORDING	t u e	**TABLE** • Title – above, in capitals, underlined (1) • Column & row headings(with units) (1) • Enclosed and neat (1)	t u p **GRAPH** • Title below, capitals, underlined (1) • Both axes labelled with units (1) • Accurate plots (1)	3	
R – REPORTING	a r g s	• Aim in capital letters (1) • Reflection appropriate (1) • Acceptable language and expression – subject-verb agreement/ grammar (1) • Spelling correct throughout with 0-3 errors (1) *Note: > 3 grammatical errors = 0 marks. AND > 3 spelling errors = 0 marks*		4	
TOTAL				10	

NAME:_____ CLASS:_____ SKILL: <u>DR</u>

TOPIC: GROWTH   SYLLABUS OBJECTIVE: B 4.12, 8.1, 8.2

17.   INVESTIGATING THE STRUCTURE OF A DICOTYLEDONOUS SEED

AIM:_____

_____

Seeds contain the next generation of plants. You will become familiar with the internal and external parts of a seed being able to identify the embryo, seed coat and cotyledons.

APPARATUS AND MATERIALS:

- Seeds – red bean (*Phaseolus multiflorus*) or other suitable dicot.
- Water

- Magnifying glass
- White tile

DICOTYLEDONOUS BEAN SEED

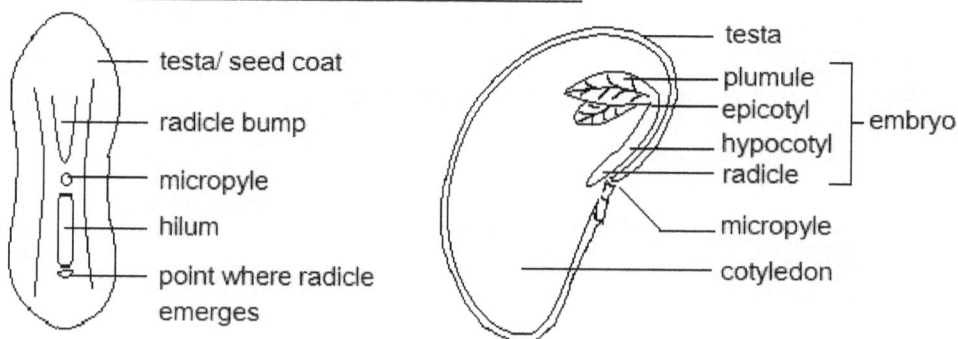

A) EXTERNAL FEATURES
(VENTRAL VIEW)

B) INTERNAL FEATURES
(LATERAL VIEW)

INSTRUCTIONS:

1. Examine a red bean seed. Notice the external features with a magnifying glass.
2. Carefully peel off the testa, and split the seed into two, being careful so the embryo is not knocked out of it. If necessary, soak the seed in some water.
3. Examine the internal features of both halves of the seed.
4. Draw and label the internal and external features of the seed.
5. Calculate the magnification for each drawing.

    Rewrite your method in past tense in the space below or on a separate page.

METHOD:

_____
_____
_____
_____

RESULTS:

Record in each box below the EXTERNAL and INTERNAL appearances of the seed. ANNOTATE at least 3 labels in each diagram. **Use additional paper as needed**.

CALCULATION OF MAGNIFICATION:

$$\frac{magnification}{of\ drawing} = \frac{size\ of\ drawing}{size\ of\ specimen}$$

$$= \underline{\hspace{2cm}} / \underline{\hspace{1.5cm}}$$

$$= \underline{X\hspace{2cm}}$$

TITLE:_____

DRAWING CHECKLIST
CLARITY: large
clean/ smooth
no shading
ACCURACY specimen proportion
LABELLING parallel
accurate lower case
justified
annotations
magnification
title & view
TOTAL

CALCULATION OF MAGNIFICATION:

$$\frac{magnification}{of\ drawing} = \frac{size\ of\ drawing}{size\ of\ specimen}$$

$$= \underline{\hspace{2cm}} / \underline{\hspace{1.5cm}}$$

$$= \underline{X\hspace{2cm}}$$

TITLE:_____

DRAWING CHECKLIST
CLARITY: large
clean/ smooth
no shading
ACCURACY specimen proportion
LABELLING parallel
accurate lower case
justified
annotations
magnification
title & view
TOTAL

NAME:_____**CLASS:**_____   **SKILL: DR**____

TOPIC: GROWTH   SYLLABUS OBJECTIVE: B 4.12, 4.13, 6.1, 8.1

18.   INVESTIGATING GROWTH AND STAGES OF GERMINATION IN RED BEAN SEEDS

AIM:_____

_____

Many plants begin their life as a seed. When seeds germinate, life begins. This experiment requires you to observe and record the transformation of a seed through the growth of the radicle to form a new seedling plant.

APPARATUS AND MATERIALS:

- Seeds – red bean (*Phaseolus multiflorus)* at different stages of growth/ germination*
- 5 Beakers/ Petri dishes
- Cotton/ tissue paper
- Water
- Magnifying glass

- White tile
- Labels
- * Alternative students each follow growth of ONE seed over a week.

INSTRUCTIONS:

1. Obtain red bean seeds, on day 0, soak approximately 5 in a beaker on some moist cotton or tissue paper, label the beaker with the date.
2. Leave the beans to germinate.
3. On day 3 place an additional 5 beans to soak and germinate. Also label the beaker with the date.
4. Repeat steps 1 -2 for two more sets of red bean seeds, ensuring the tissue paper remains moist for the duration of the lab experiment.
5. After 1 - 2 weeks, there should be red-bean seeds at various stages of germination.
6. Draw and label the stages of germination to represent the seed/ seedling on day 0, day__, day__ and day___.

   Rewrite your method in the past tense in the space below or on a separate page.

METHOD:

_____
_____
_____
_____
_____
_____
_____
_____
_____

RESULTS:

*Record in each box below the appearance of the seed/ seedling on the various days. **Use additional paper as needed**.*

DAY 0-1

CALCULATION OF MAGNIFICATION:

$$\frac{magnification}{of\ drawing} = \frac{size\ of\ drawing}{size\ of\ specimen}$$

$$= \underline{\hspace{1cm}} / \underline{\hspace{1cm}}$$

$$= \ X \ \underline{\hspace{2cm}}$$

DRAWING CHECKLIST

CLARITY:
large

clean/ smooth

no shading

ACCURACY
specimen proportion

LABELLING
parallel

accurate
lower case

justified

annotations

magnification
title & view

TOTAL

TITLE:_____

DAY __

CALCULATION OF MAGNIFICATION:

$$\frac{magnification}{of\ drawing} = \frac{size\ of\ drawing}{size\ of\ specimen}$$

$$= \underline{\hspace{1cm}} / \underline{\hspace{1cm}}$$

$$= \ X \ \underline{\hspace{2cm}}$$

DRAWING CHECKLIST

CLARITY:
large

clean/ smooth

no shading

ACCURACY
specimen proportion

LABELLING
parallel

accurate
lower case

justified

annotations

magnification
title & view

TOTAL

TITLE:_____

DAY __

DRAWING CHECKLIST

CLARITY:
large

clean/
smooth

no shading

ACCURACY
specimen
proportion

LABELLING
parallel

accurate
lower case

justified

annotations

magnification
title & view

TOTAL

CALCULATION OF MAGNIFICATION:

$$\frac{magnification}{of\ drawing} = \frac{size\ of\ drawing}{size\ of\ specimen}$$

$$= \underline{\quad\quad}/\underline{\quad\quad}$$

$$= \underline{X\quad\quad\quad}$$

TITLE:_____

REFLECTION _____

_____

_____

_____

DISCUSSION QUESTIONS:

1. Explain germination, identifying 3 environmental/ abiotic factors needed.
2. After soaking for 24 hours, the radicle of the seed begins to emerge first. What changes take place inside the seed to cause the growth of the radicle?
3. Why does the radicle emerge first?
4. How and why do the cotyledons change over time?
5. Why were some of the structures observed green – hypocotyl, epicotyl and cotyledons?

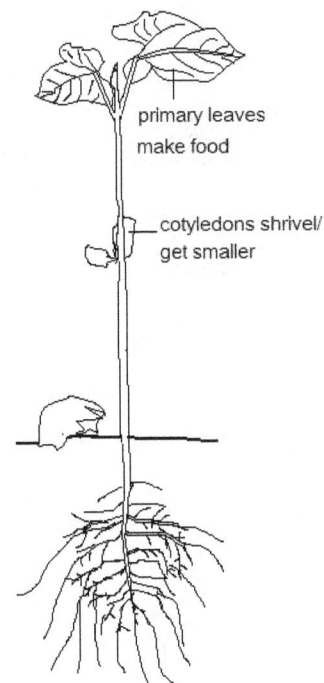

NOTES - GERMINATION

Germination describes the growth of the seed into a seedling. The seed contains the embryo - made up of the plumule and the radicle and food stored in the cotyledon and/or endosperm. The embryo is protected by the tough testa (seed coat).

In its inactive and dehydrated state, a seed can stay a seed for a long time. It is said to be dormant. When conditions are favourable, germination begins. Germination Stages: 1) Seed absorbs water (imbibition), swells and testa splits open. Stage 2) Respiration in cells, and Stage 3) Food moves to growing points/ meristems.

Germination requires 3 abiotic conditions:

1. **Water** – moves rapidly into the micropyle and to all the cells to activate enzymes. Amylase enzyme causes starch to break down to glucose to be used in respiration. Proteases also breakdown protein into amino acids. These breakdown products are used to form new cells in the radicle and plumule, causing them to grow.
2. **Oxygen** – needed for respiration of the glucose (to provide energy to form new structures)
3. **Warmth/ suitable temperature** –the optimum temperature for the enzymes to work efficiently.

Cotyledons emerge from the testa during germination. Over time, they become smaller (mass decreases) as stored food is converted and used to grow the new seedling. In red bean plants the cotyledons move above ground (epigeal germination). They are usually green, photosynthesizing until the first foliage leaves unfurl, then they drop off.

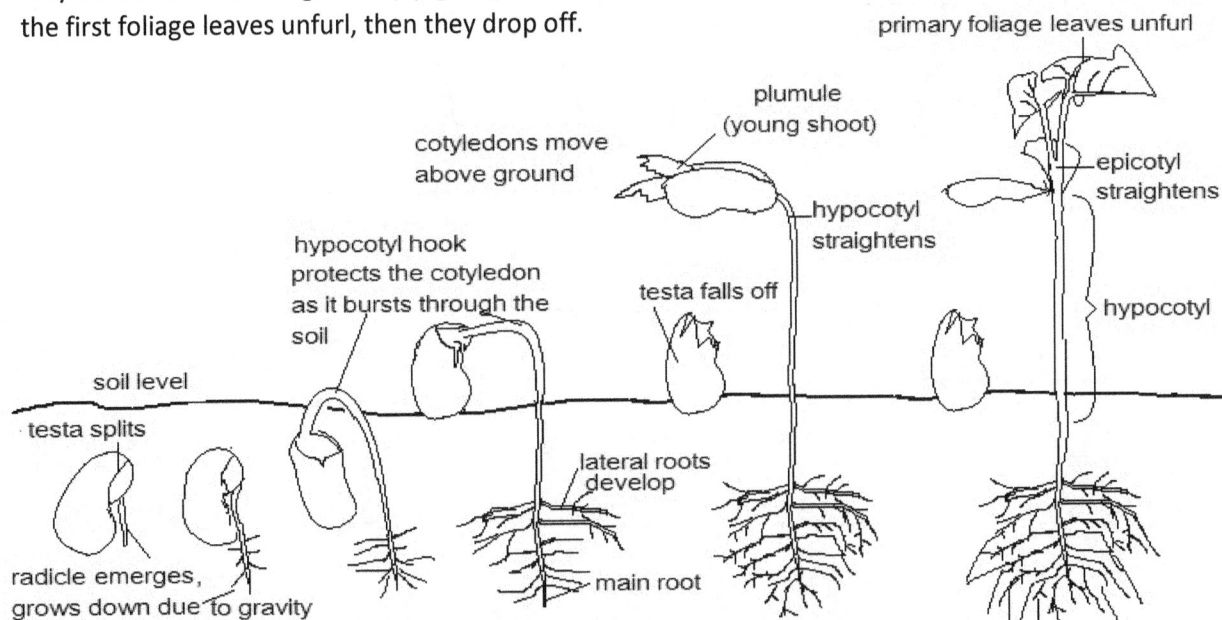

DRAWING SHOWING STAGES OF EPIGEAL GERMINATION (COTYLEDONS COME ABOVE GROUND)

NAME:_____CLASS:_____   SKILL: MM/ORR

TOPIC: GROWTH   SYLLABUS OBJECTIVE: B 4.12, 4.13, 6.1, 8.1

19.   INVESTIGATION TO DETERMINE IF OXYGEN IS NEEDED FOR GERMINATION

AIM:_____

_____

In this experiment you are required to investigate if oxygen is needed for the process of germination in bean seeds.

APPARATUS and MATERIALS:

- 6-8 Mung bean seeds
- Cotton wool
- Thread
- Scissors
- 1 Spatula

- 50cm^3 Measuring cylinder
- 2 Conical flasks with rubber bungs
- Labels

- Sodium hydroxide (NaOH) solution
- Pyrogallic acid
- Distilled water

DIAGRAM:

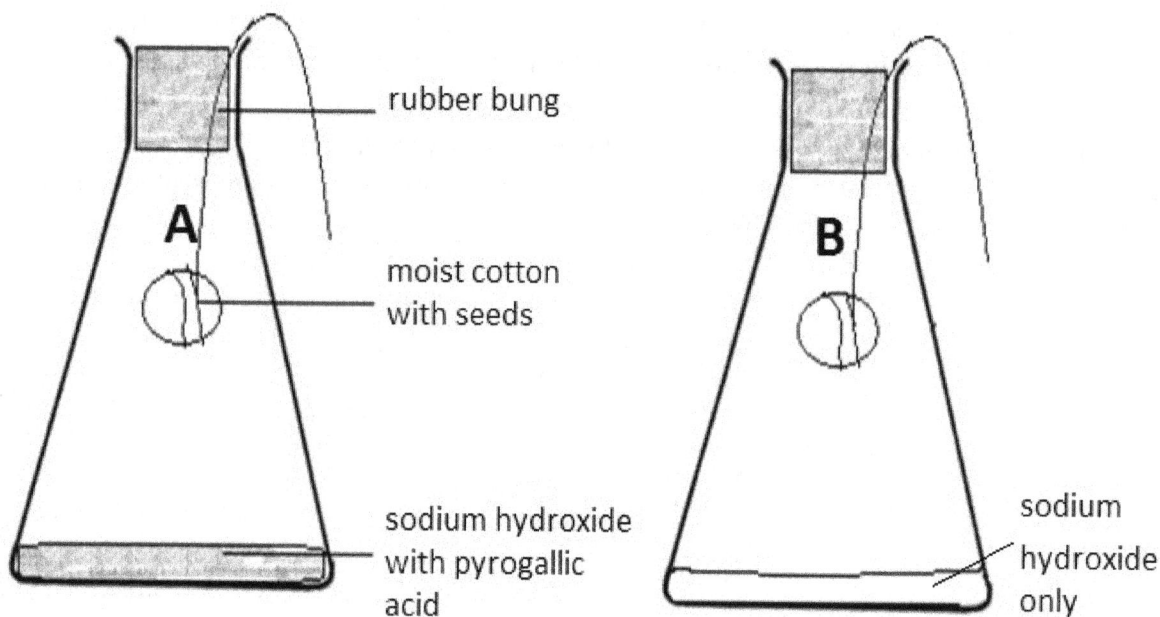

DRAWING SHOWING SET-UP OF INVESTIGATION IF OXYGEN IS NEEDED FOR GERMINATION

INSTRUCTIONS:

1. Label two conical flasks A and B, measure 25cm^3 sodium hydroxide solution and place into each.

2. Moisten two pieces of cotton wool and roll and insert 4 mung beans in each. Ensure the seeds cannot fall out and use thread to secure the cotton ball.

3. Measure a piece of thread to tie each piece of cotton wool to a bung so that the cotton ball will hang above the solution in the flask as shown in the diagram.

4. Add 2 spatulas full of pyrogallic acid to the conical flask A. Cover immediately with the bung, then swirl the flask to mix. The solution will turn dark brown or black.

5. Take one piece of cotton ball with the beans and quickly suspend in flask A. Cover immediately. ***Do not let the cotton touch the sides of the flask or the solution in the flask**. Do the same with flask B.

6. Leave both flasks in a warm place for 48 hours.

7. At the end of two days, note how many seeds germinated in each flasks in an appropriate table.

Rewrite your method in past tense in the space below or on a separate page.

METHOD:

_____

_____

_____

_____

_____

_____

_____

_____

_____

_____

_____

_____

RESULTS: (*Add a title to the table and record the observations.*)

TABLE SHOWING _____

Time (Days)	Description of Flasks	
	Flask A: pyrogallic acid and sodium hydroxide Oxygen absent	Flask B: sodium hydroxide only Oxygen present
0		
1		
2		
	Inference:	Inference:

Include a drawing of the cotton balls at the end of the experiment for both conditions. Add annotations to describe the observations.

TITLE:_____

DISCUSSION

_____

_____

_____

_____

_____

_____

_____

_____

_____

_____

_____

_____

_____

_____

_____

_____

LIMITATIONS

_____

PRECAUTIONS

_____

SOURCES OF ERROR

_____

CONCLUSION (*Was oxygen needed for germination? What do the results show?*)

_____

_____

_____

REFLECTION (*Do you have a better appreciation/ understanding for germination requirements?*)

_____

_____

_____

DISCUSSION QUESTIONS:

1. What is germination? What are the conditions needed for germination?
2. Identify the control set-up in this experiment.
3. Describe the conditions and the results obtained in set-up A. Was this result as expected? Include the function of NaOH and pyrogallic acid.
4. Describe the conditions and the results obtained in set-up B. Was this result as expected? Why was only NaOH added?
5. According to the results, is oxygen needed for germination? Why? (Hint: what specific cellular processes requires oxygen?)
6. How can this experiment be improved?
7. Explain why the following precautions were taken while doing the lab:
   - Carefully adding acid
   - Closing the flasks quickly
   - Placing the flasks in a warm place
   - Ensuring no acid or solution got on the cotton ball with seeds
8. What are some limitations?
   a. Seeds did not germinate because they were old
   b. The temperature was not warm enough

NOTES - GERMINATION

Germination is the process whereby a seed begins to form new life. The new life becomes visible when the radicle emerging through the seed coat. In general, the conditions needed for germination are oxygen, water and warm temperatures. First, **water** is required for chemical reactions to begin such as the **activation of enzymes** followed by breakdown of stored food into soluble substance. In order to breakdown these substances, **oxygen** is required for aerobic respiration to **generate energy**. **Warm temperatures** provide optimum conditions for these cellular activities to occur.

Note: When pyrogallic acid is mixed with sodium hydroxide, it absorbs oxygen and carbon dioxide and should be in an alkaline solution. The function of NaOH is to absorb CO_2 and create an alkaline environment for pyrogallic acid.

Critera – Manipulation and Measurement To see if oxygen is needed for germination (pyrogallic acid)	MM Marks	
• Correct temporary storage of stopper upside down	1	
• Stopper replaced immediately after use of reagent	1	
• Measurement of NaOH – reading measuring cylinder at eye level	1	
• Measuring pyrogallic acid – 1 spatula, no spills	1	
• Careful adding pyrogallic acid in flask and swirling to mix	1	
• Wrapping of seed securely with cotton	1	
• Tying thread – so cotton does not fall into the mixture	1	
• Correct set up of apparatus – replace bung quickly	1	
• Labelling of samples	1	
• Clean up work station	1	
TOTAL	**10**	

NAME:_____**CLASS:**_____ **SKILL: ORR**__

TOPIC: IRRITABILITY   SYLLABUS OBJECTIVE: B: 7.6

20.   INVESTIGATING REACTION TIMES OF STUDENTS CATCHING A FALLING METER RULER

AIM:_____

_____

How fast do you think you are? Do you know what a reflex and a reaction are? For which one do you have to think before an action is taken? This investigation tests your reaction time by measuring how quickly you can catch a falling meter ruler. Given different stimuli – sight and touch, sight alone or touch alone, you can compare your reaction times using different stimuli and even against other students.

APPARATUS AND MATERIALS:
- A meter ruler
- 2 students
- Blind fold (optional)
- calculator

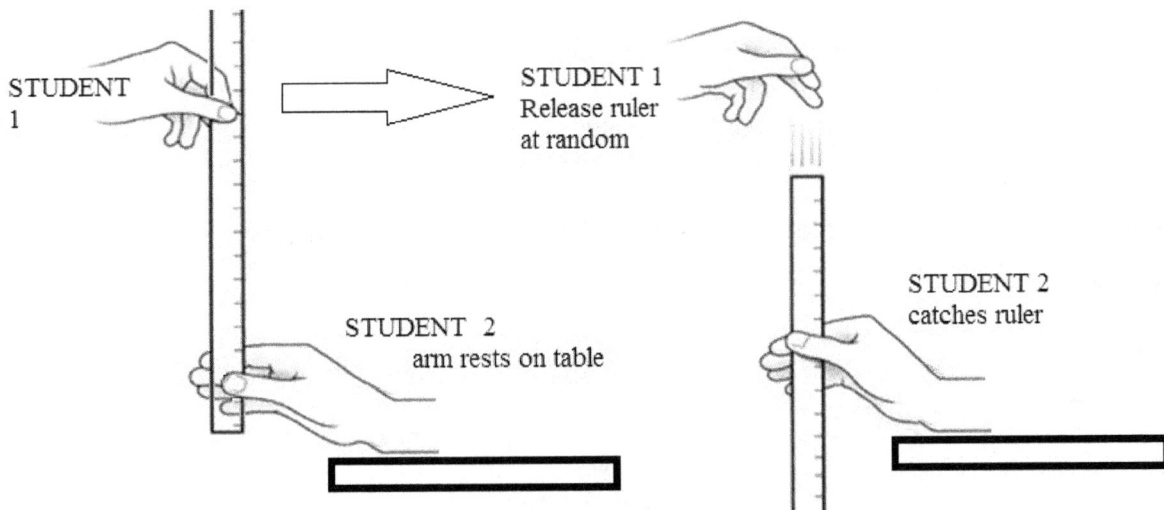

DIAGRAM SHOWING SET-UP OF STUDENT 1 AND 2 TO INVESTIGATE REACTION TIMES OF STUDENTS

INSTRUCTIONS:

1. Work in pairs. Student 1 stands and drops the ruler, while student 2 has to catch it between the thumb and forefinger of his/her dominant hand while sitting near the edge of the table, resting his/her elbow and forearm and leaving his/her wrist hanging over the side.
2. With student 1 holding the ruler vertically, allow student 2 to see and touch the ruler at the 20cm mark. This is the first scenario using two stimuli – **sight and touch**.
3. Student 1 waits a few seconds then allows the ruler to fall at random while student 2 catches the ruler <u>as quickly as he/she can</u>. Student 1 must not say when he/she drops the ruler.
4. Record the distance the ruler has fallen (by subtracting 20cm from the measurement on the ruler where student 2 caught it).
5. Repeat the dropping and catching using both sight and touch for two more trials.
6. Now repeat steps 1 - 4, with the ruler visible to student 2, but not touching his/her fingers. This is the second scenario, with one stimulus – **sight only**.
7. Finally repeat steps 1 - 4, with student 2 blindfolded or eyes closed but the ruler touching his/her fingers. This time the stimulus is **touch only**.
8. Calculate the average (mean) distance the ruler has fallen under the three different scenarios and stimuli. Convert the distances in centimetres to distances in meters. Use 2 d.p.
9. Convert distance to reaction time using the formula:

$$\text{Reaction time} = \sqrt{\left(\frac{2d}{g}\right)}$$

Where: d = the distance the ruler fell **in meters**, g = the acceleration of gravity (9.8 m/s^2)

10. Plot a bar graph to compare the average reaction times of students when catching a falling meter ruler using different stimuli.
11. *Optional:* Repeat the entire experiment with student 2 dropping the ruler and student 1 catching given the three scenarios.

Rewrite your method in past tense in the space below or on a separate page.

METHOD:

_____

_____

_____

_____

_____

_____

_____

_____

_____

_____

_____

RESULTS: *(Complete the table below by filling in the columns; and adding a title)*

TITLE: _____

Name of student	Scenario/ stimulus	Distance the ruler dropped (cm)				Average distance ruler dropped (m) (d) (100 cm = 1m)	Reaction time $\sqrt{\left(\dfrac{2d}{g}\right)}$	Notes: Hand used (athletic/age/etc.)
		Trial 1	Trial 2	Trial 3	Average			
A. Student 2 catching	Sight and Touch							
	Sight only							
	Touch only							
B. – Optional – student 1 catching	Sight and Touch							
	Sight only							
	Touch only							

Plot a graph below using the checklist:
□ *Title*   □ *Scale*   □ *Labels on both axes*   □ *Units in each label*   □ *Plots correct*   □ *Line*

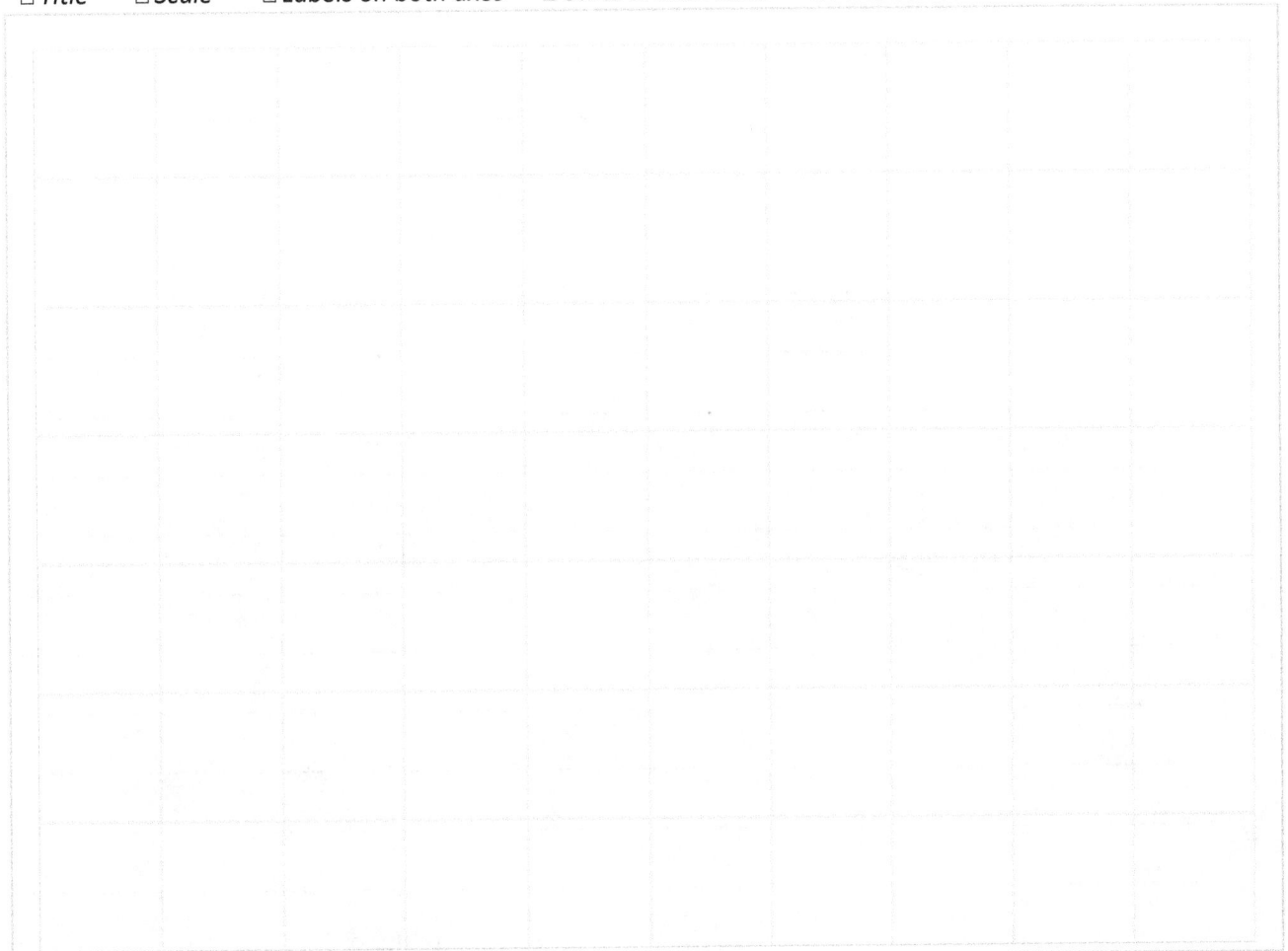

TITLE: _____.

DISCUSSION

LIMITATIONS _____

_____

_____

PRECAUTIONS _____

_____

_____

SOURCES OF ERROR _____

_____

_____

CONCLUSION – *When are reaction times fastest or slowest? Which sense(s) are used?*

_____

_____

_____

REFLECTION – *What do you have a better appreciation for or understanding about on this topic?*

_____

_____

_____

_____

_____

SUGGESTED MARK SCHEME - ORR CRITERIA			Mark
O – **Observations**	s c/d	• Significant changes noted –notes recorded in the table (1) • Complete data set in columns 3 – 5 of the table of results (1)	2
RTG – **Recording**	t u p	**GRAPH** • Title below, capitals, underlined (1) • Both axes labelled with units (1) • Accurate plots (1) • Bars not touching – bar chart/ discontinuous data (1)	4
R – Reporting	a r g s	• Aim in capital letters (1) • Reflection appropriate (1) • Acceptable language and expression – subject-verb agreement/ grammar (1) • Spelling correct throughout with 0-3 errors (1) *Note: > 3 grammatical errors = 0 marks. AND > 3 spelling errors = 0 marks*	4
TOTAL			10

DISCUSSION QUESTIONS

1. What is the difference between reflex actions (reflexes) and a voluntary/ conscious action (reaction)? Which one was examined in this experiment?

2. From the results, identify the stimulus that caused the fastest reaction times. Quote values from the graph. (Note: fastest reaction times are the lowest bar on the chart; faster reactions mean the ruler fell a shorter distance.)

3. Was this expected? What are possible reason(s) for the results varying from the expected?

4. List all the parts of the nervous system through which impulses passed for:
   a. Touch stimulus b. Sight stimulus

5. Will the students' reaction times improve with practice/ more trials? Why? (Hint: more practice allows for conditioned responses or muscular memory.)

6. Compare the reaction times of various members of the class. Is there any correlation with sporting ability, age, height, longer limbs – longer distance for impulses to travel?

7. Identify at least 2 limitations/ sources of error in this experiment. (Hint: health or tiredness of a person may not have been taken into account.)

NOTES: - RESPONSES TO STIMULI

Every day you have reactions/ responses to stimuli. How quickly you can respond may be rewarding, like catching a cricket ball. Slower responses may have disastrous consequences, such as getting hit in the face with a football. Your reaction time is a measure of how quickly you can respond to a stimulus. **Reflexes are rapid, automatic, involuntary responses** to a particular stimulus important for survival. A reflex does not involve conscious decision making in the brain; but a message will be sent to your brain (soon after). Usually reflexes are faster than other voluntary actions/ responses. **Voluntary actions are consciously controlled**; after processing and decision making in the brain, a reaction occurs.

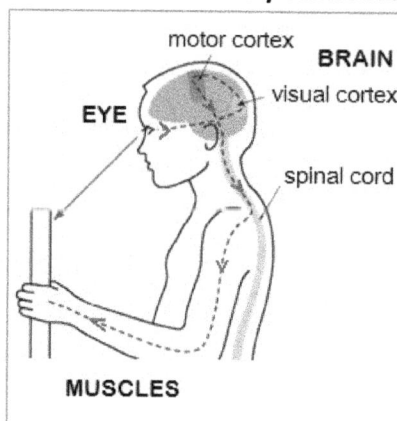

The reaction pathway for **Sight Stimulus:**

- **Stimulus** - **The eye** sees the ruler drop.
- The eye sends a message to the visual cortex via the optic nerve – **a sensory neurone**.
- **In the brain** - The visual cortex sends a message to the motor cortex which sends a message to the spinal cord.
- The spinal cord sends a message to the hand/finger muscle along a **motor neurone**.
- **Response** - The **finger muscle** contracts to catch the ruler.

Some reactions have to be learnt first, then they can be performed without conscious thought. When we acquire a new physical skill through repetition, our nervous system creates new neural pathways. More practice catching the ruler causes the eye, brain and muscles to become better connected and efficient. This connection creates **muscle memory (conditioned response)**. However, no matter how good the muscle memory becomes, it will always take some time for a response to occur.

NAME:_____**CLASS:**_____ **SKILL: ORR_____**

TOPIC: IRRITABILITY   SYLLABUS OBJECTIVE: B: 7.6, 7.9

21.   INVESTIGATING THE EFFECT OF LIGHT INTENSITY ON THE PUPIL SIZE

AIM:_____

_____

Do you have control over all of your body's actions? What happens when a bright light is shone in your eyes? In this experiment, you will determine the effect of light on the size of your pupils.

APPARATUS AND MATERIALS:

- 2 Students
- Mirror
- Torchlight/ light source
- Meter rule
- Ruler

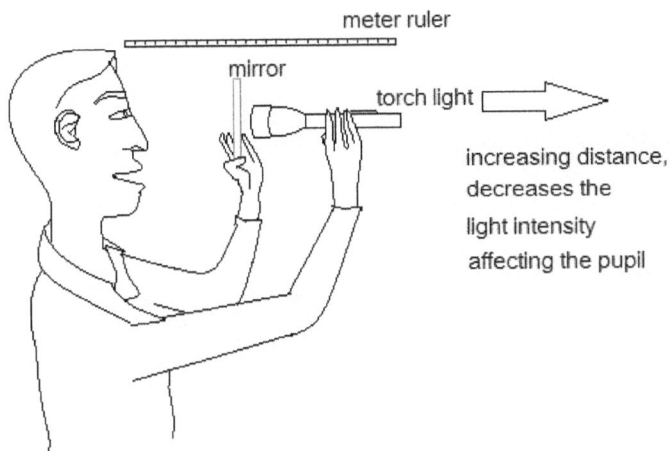

meter ruler

mirror

torch light

increasing distance, decreases the light intensity affecting the pupil

DRAWING SHOWING SET-UP OF EXPERIMENT TO INVESTIGATE PUPIL SIZE

METHOD:

1. Leave the overhead lights on in the room, estimate the distance that those bulbs are away from the student experimenter's eyes. Record the distance and consider this low light intensity.
2. Examine the eyes using a mirror. Notice the white of the eyes, the colour of the iris and the pupil (a black hole at the centre of the iris).
3. Using the ruler, carefully measure the size of the pupil on the mirror.
4. Holding the torchlight at the same distance as the mirror, shine the light into the eye. Try to measure the new size of the pupil. Record this as high light intensity. Also measure the distance between the torchlight and the eye.
5. Notice that it takes longer for pupils to dilate (widen) than it takes to contract. Sometimes, too, the pupil gets too small and has to re-open slightly.
6. With the assistance of a student, hold the torchlight, further away from the eyes, but not as far as the light bulb; measure and record the distance from the eye as well as the new diameter of the pupil in the mirror.

RESULTS:

Fill in the table below with the measurements taken

TABLE SHOWING THE EFFECT OF LIGHT INTENSITY ON THE PUPIL OF THE EYE

Distance of light source from eye (cm)	Light Intensity (Description)	Diameter of Pupil (mm)
	Low	
	Medium	
	High	
	Very high	

In the space below add a drawing to show the diameter of the pupil in low light and in very bright light.

DISCUSSION

_____

_____

_____

_____

_____

_____

_____

_____

_____

_____

_____

_____

_____

_____

_____

_____

_____

LIMITATIONS

_____

_____

PRECAUTIONS

_____

_____

SOURCES OF ERROR

_____

_____

CONCLUSION (*What is the effect of light intensity, as represented by distance, on pupil size?*)

_____

_____

REFLECTION – *What impact did this experiment have on your understanding of reflex actions?*

_____

_____

_____

DISCUSSION QUESTIONS:

1. Identify the parts of the eye involved in controlling the pupil response (stimulus, receptor, effector and response in bright light).
2. Explain how the pupil's response to bright light is important to the body.

3. Based on the results, what effect does increasing light intensity have on the pupil?

4. Identify limitations/ difficulties in doing this experiment. (Hint: difficult to measure pupil diameter on the mirror, eyes become tired, diseased eye/ damaged eye/ intoxicated person, shifting distance of light during measurement, different light sources should not be compared).
5. How can the experiment be improved? (Hint: repeating the experiment with the same person, repeating with different persons, using a magnifying glass in front of the mirror)

NOTES:

The pupil of the eye is an opening that allows light onto the retina. It appears black and is surrounded by the coloured part of the eye, the iris. Pupils range in diameter from 1.5mm to 8mm, getting wider/ expanding to allow more light to enter the eye. When light stimulus falls on the receptors in the retina, it sends nerve impulses directly to the muscles of the iris (effector), signally them to contract or relax (response). Both pupils are controlled together as part of the optic nerve from one eye couples with muscles that control the other pupil size.

Pupillary reflex/ Iris-Pupil reflex (Involuntary reflex)
Control of pupil size is an involuntary, cranial reflex action. It involves contraction and relaxation of the antagonistic muscles of the iris (effectors) in response to changes in light intensity (stimulus). A narrow pupil at high light intensities prevents overstimulation damage to the photoreceptors in the retina, while at low light intensities, the pupil dilates so that a clear image can be formed.

The size of your pupil can also indicate your emotions – such as fear, anger, pain, love or the influence of drugs – in this case, emotions stimulate the iris muscles. A lack of a pupil reflex can indicate nerve damage or brain death. This is why doctors flash lights into persons with suspected head injury or brain impairment (alcohol overdose, etc).

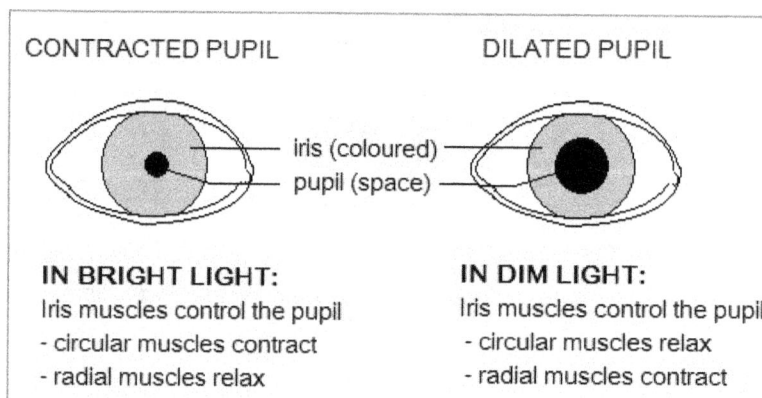

CONTRACTED PUPIL   DILATED PUPIL

iris (coloured)
pupil (space)

IN BRIGHT LIGHT:
Iris muscles control the pupil
- circular muscles contract
- radial muscles relax

IN DIM LIGHT:
Iris muscles control the pupil
- circular muscles relax
- radial muscles contract

NAME:_____ CLASS:_____ SKILL: AI____

TOPIC: IRRITABILITY & GROWTH   SYLLABUS OBJECTIVE: B: 7.1, 7.2, 6.1, 4.13, 8.1

22.   INVESTIGATING THE EFFECT OF UNIDIRECTIONAL SOURCE OF LIGHT ON PLANT GROWTH

AIM:_____

_____

In this experiment, you are expected to measure height as an indication of growth of seedlings and also investigate the effect of a unidirectional light source on overall plant growth. You must be very responsible and careful to measure your plants every day and to water them.

APPARATUS AND MATERIALS:

- 2 **Transparent plastic** cups (8oz.) with drainage holes
- Moist tissue
- Promix Soil – enough for each cup
- Water
- String (approx. 60cm)

- Ruler
- Labels
- Red bean seedlings (germinated from seed)*
- Dark box with a slit cut on one end
- Trolley

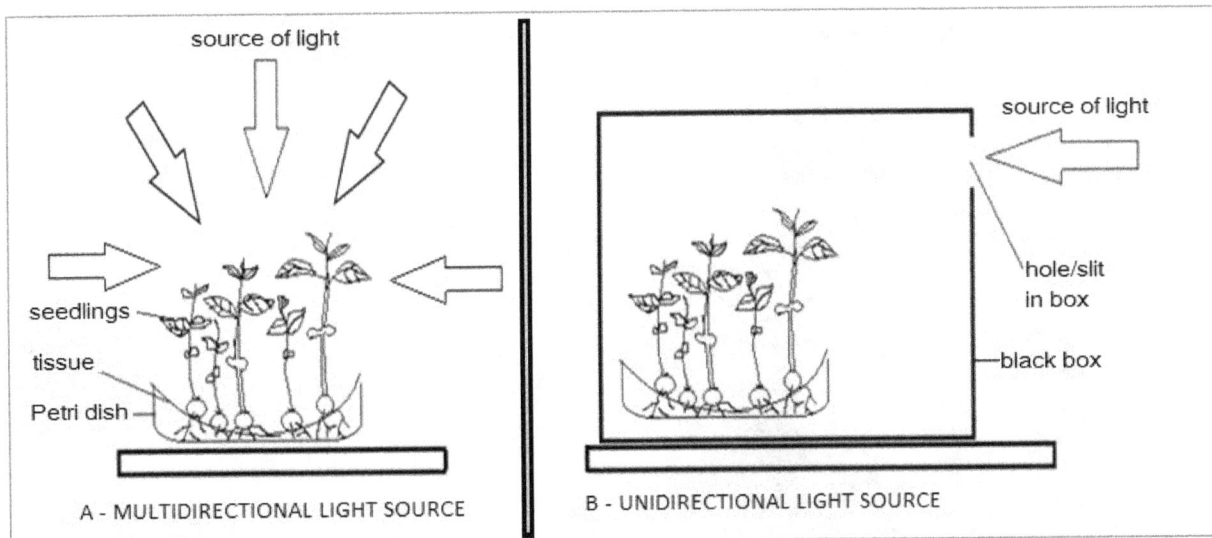

SET-UP OF SEEDLINGS IN MULTIDIRECTIONAL AND UNIDIRECTIONAL SOURCES OF LIGHT

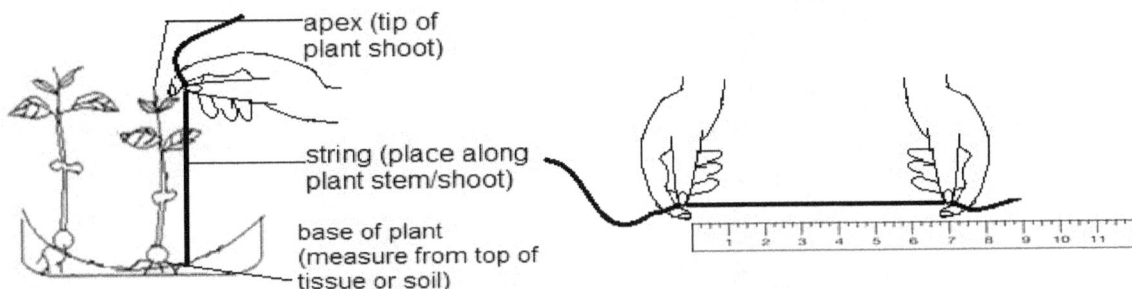

DRAWING SHOWING HOW TO MEASURE THE HEIGHT OF THE SEEDLINGS USING STRING AND A RULER

INSTRUCTIONS:

1. Obtain two transparent plastic cups with small drainage holes at the base.
2. Place 3 red bean seeds up against the side of each the transparent plastic cup using a wet tissue paper. Back-fill the cup with the promix soil until it is ¾ filled.
3. Allow the seeds to germinate by leaving them in a warm place for two to three days. Do not allow the soil to dry out.
4. Label the cups A – multidirectional/full light and the other B – unidirectional light, and include the name of your group.
5. In experiment A and B, label the three seedlings 1, 2 and 3 respectively.
6. Place cup A on the lab bench, turning it a few times every day so that light comes from all direction.
7. Place seedling B inside a large box, as far away from the slit on one side. Do not turn it at all.
8. Every day, using a string and ruler, carefully measure the heights of the seedling shoots from the seed to the growing tip above.
9. Record the general direction of each seedling under both conditions.
10. Return seedlings B to the same spot and same orientation every day.
11. Leave both set-ups for about 1 week, making sure to water each day.
12. Draw the seedlings in each cup at the end of the experiment.
13. Plot a graph to show the average change in height of the seedlings over time.

Rewrite your method in past tense in the space below or on a separate page.

METHOD:

RESULTS: *A – **Record in each box below the heights** and appearance of the seedlings every day.*

TABLE SHOWING THE HEIGHTS AND APPEARANCE OF THE SEEDLINGS GROWING IN MULTI- AND UNI-DIRECTIONAL SOURCES OF LIGHT

| Time (Days) | Height of seedlings (cm) | | | | | | | | | |
| | A – Multidirectional light source | | | | | B – Unidirectional light source | | | | |
	1	2	3	Avg	Seedling descriptions	1	2	3	Avg	Seedling descriptions
0										
1										
2										
3										
4										
5										
6										
7										
8										
9										
10										

*B - **Draw a sketch** of the seedlings at the end of the experiment. Add descriptions – shape and colours.*

Description of plant: _____

Colours of stems/leaves: _____

Description of plant: _____

Colours of stems/leaves: _____

TITLE:_____

C - On a graph page **plot a graph** *to show the change in height of each seedling over the 10 day period.*
□ Title   □ Scale   □ Labels on both axes   □ Units in each label   □ Plots correct   □ Line

TITLE OF GRAPH :

DISCUSSION

LIMITATIONS _____

_____

_____

_____

_____

PRECAUTIONS _____

_____

_____

SOURCES OF ERROR _____

_____

_____

_____

CONCLUSION:

_____

_____

_____

_____

_____

_____

REFLECTION: *How is the relationship between growth and light important to you ? What was most significant? How did you feel about the duration of lab, working with live specimens?*

_____

_____

_____

_____

_____

_____

DISCUSSION QUESTIONS:

1. Define growth, stimulus and response in relation to this experiment.
2. Explain which part of the shoot was sensitive to light.
3. Why is it important for the plant shoots to bend towards the light? Explain how auxins cause phototropisms in plant shoots.
4. In what direction was the seedlings expected to grow with a unidirectional light source (B)? Did this occur and why/why not? Refer to the table or drawings in the results using data.
5. How were the seedlings receiving light from multidirectional sources (A) expected to grow? Do the actual results confirm this? If not why?
6. **Describe** the shape/ trend of the graph and **identify** which set of seedlings grew taller/longer. **Explain** why this occurred, **quoting** the maximum heights for each treatment from the graph.
7. What is etiolation? Which set of seedlings showed this and why?
8. Identify the control in this experiment.
9. How can the method be improved?
10. What precautions were taken? (Completely darkening the box, small singular hole, watering the seedlings regularly, being gentle, etc.)
11. What limitations or factors were difficult to control? (Limited light due to cloudy weather, exposure to light while measuring "B")

NOTES - TROPISMS – THE DIRECTION OF PLANT GROWTH

Plants need light and water for **photosynthesis**. **Tropisms** ensure they grow towards sources of light and water.

- **Positive tropism** – towards the stimulus
- **Negative tropism** – away from the stimulus
- **Phototropism** – growth response to the direction of **light**
- **Geotropism** – growth response to the direction of **gravity**

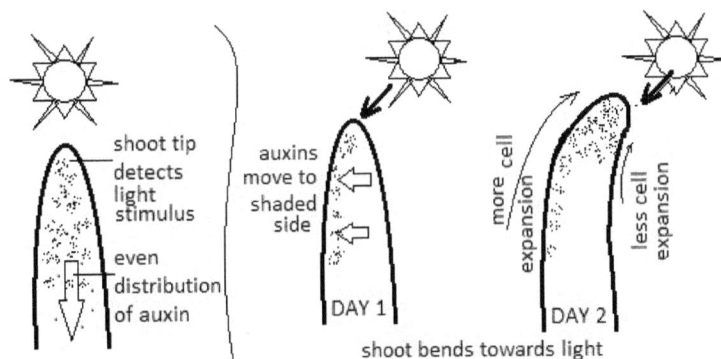

shoot tip detects light stimulus

even distribution of auxin

auxins move to shaded side

more cell expansion

less cell expansion

DAY 1   DAY 2

shoot bends towards light

Responses of different parts of the plant:

Response	Part of plant	Direction of growth	Advantage
Positive phototropism	Shoot tip	Growth towards light	Allows maximum light for photosynthesis
Negative phototropism	Root tip	Growth away from light	Less chance of drying out

Auxin is a plant hormone that controls the direction of growth of root tips and shoot tips in response to **different stimuli** - light and gravity. Auxin is made at the **tips** of shoots. It is moved in solution down to older parts of the shoot. It causes cells to **grow longer.** Light causes uneven distribution of the auxins, accumulating on the shaded side of shoots **causing bending** in the shoot tip.

ETIOLATION: occurs in plants **grown in partial or complete absence of light**. It is an **adaptation/ response** that is characterized by: **long, weak stems;** smaller unexpanded leaves due to longer internodes; and a **pale yellow colour (chlorosis).**

CSEC Sample mark scheme for AI skill – To investigate light on germination of seedlings

AI GENERAL CRITERIA			SPECIFIC CRITERIA –	MAX MARK
B-Background	d s	• Define key related terms (1) • Statement of relevant theory (1)	- What are tropisms?, Light stimulus identified - How do auxins cause bending of tips.	2
E - Explanation	T C u	- Trends and patterns identified (1)	- Describe shape of graph for plant with multidirectional light AND unidirectional light. Identify which grew taller/ faster - Etiolation explained if observed and why.	4
	m	- Compare actual results with expected results (1) - Use data to support explanations (1) - Modification/ improvement to existing method (1)	- Expected results stated and compared with actual results - Values from the graph quoted correctly - Identify various soils, other plant pots, choice of room vs outside, etc.	
LSP - Limitations Sources of error/ Precautions		- Identify at least 2 limitations **with explanations** - Identify at least 2 precautions/ sources of error **with explanations**	- How exposure to light while measuring may affect results – explained. OR OTHER limitation - Precautions	2
C- Conclusion	s r	- Statement - Related to aim	Unidirectional light caused shoot tips to bend/ not bend. Unidirectional light increased/ decreased length of stems over time….	2
TOTAL				10

NAME:_____ CLASS:_____ SKILL: AI____

TOPIC: IRRITABILITY   SYLLABUS OBJECTIVE: B: 7.1, 7.2b

23.   INVESTIGATING THE RESPONSE OF INVERTEBRATES TO MOISTURE AND LIGHT STIMULI

AIM:_____

_____

Can living things as small as woodlice detect and respond to changes in their environment? Do they have a preference for moisture or darkness or both? Irritability or the response to stimuli is a feature of all living organisms. It aids survival in finding food and mates, and in avoiding harsh environments that are too moist or too dry. In this experiment, you will investigate how invertebrates respond to different stimuli.

APPARATUS AND MATERIALS:

- Choice chamber apparatus OR Plastic Petri dish with clear transparent plastic cover
- Specimen jar/ beaker
- Mosquito netting/ mesh/ nylon/ muslin
- 10-12 Millipedes/ woodlice/ maggots
- Thick dark black cloth/ cardboard
- Cotton wool

- Tap water
- Silica gel/ anhydrous calcium chloride
- Water
- Labels
- Paintbrush/plastic forceps
- Tape

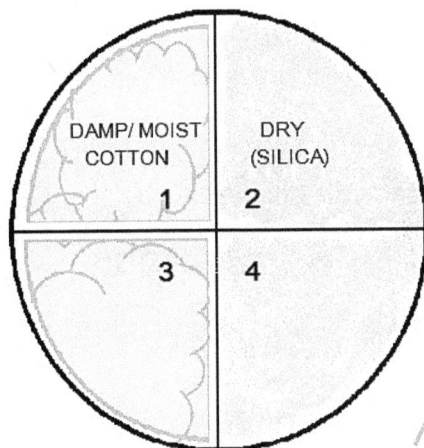

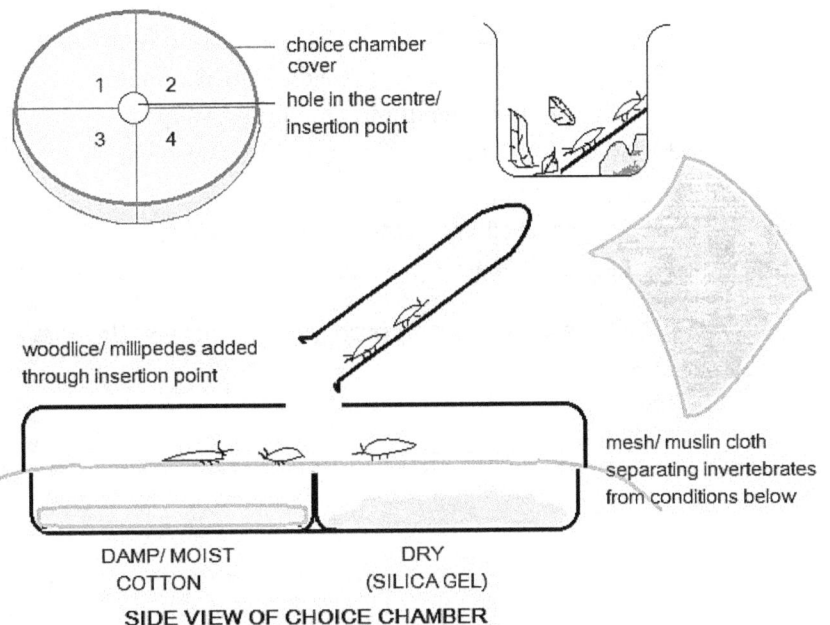

DIAGRAM SHOWING SIMPLE SET-UP OF CHOICE CHAMBER TO COMPARE RESPONSE OF INVERTEBRATES TO DAMP/ MOIST CONDITIONS VERSUS DRY CONDITIONS

INSTRUCTIONS:

1. Find woodlice (invertebrates)/ millipedes or maggots in their natural habitats by looking under rocks, digging soil or disturbing leaf litter. Carefully transfer the organisms with some of their habitat to the specimen jar using a paintbrush or plastic forceps. Be careful not to crush any organisms.
2. Obtain a choice chamber with 4 segments and label the chambers 1, 2, 3, 4.
3. Place cotton wool on one half of the lower dish (segments 1 and 3) and moisten sufficiently with tap water.
4. Place silica gel or anhydrous calcium chloride on the other half of the dish (segments 2 and 4).
5. Add the layer of mesh/ nylon/ muslin cloth over the entire bottom piece, then place the transparent cover over the top.
6. Carefully add the millipedes/ woodlice through the hole/insertion point.
7. Allow the organisms three minutes to acclimate. Then count the number of organisms on each half of the apparatus at 1 minute intervals for 5 minutes.
8. Record the number of organisms in each condition over the 5 minutes then calculate the averages. Record any other interesting behaviour of the organisms (speed of movement).
9. Plot a bar graph to show the preference of millipedes/ woodlice for moisture or dry conditions.

OPTIONAL:

10. Cover completely the outer half of the choice chamber (segments 1 and 2) by papering the bottom and top portions with black paper.
11. Leave segments 3 and 4 exposed to the light.
12. Count the number of organisms that will be introduced to the chamber. Then insert all through the hole in the cover. Use the black cloth for extra darkness on the dark sides.
13. Allow the organisms to acclimate for 3 minutes then count the number of organisms in each segment each minute for 3 minutes.
14. Work quickly when lifting the dark side cover, so that those organisms are not affected by the light and begin to move.
15. Record all counts in the appropriate table
16. Return all living organisms to their natural habitat at the end of the experiment.

METHOD: *Rewrite your method in past tense in the space below or on a separate page.*

_____

_____

_____

_____

_____

_____

_____

RESULTS: A – ***Record in each box below*** the number of organisms in each environmental condition.

TABLE 1 SHOWING THE NUMBER OF MILLIPEDES/ WOODLICE IN MOIST AND DRY CONDITIONS

Trial/ Count	Number of Millipedes/ Woodlice	
	Moist conditions	**Dry conditions**
1		
2		
3		
4		
5		
Average		

Additional observations: _____

_____

_____

*Graph – Draw **a bar chart** to compare the preference of the millipedes/ woodlice to moisture or dryness.*

Checklist: □ *Title*   □ *Scale*   □ *Labels on both axes*   □ *Units in each label*   □ *Plots correct*   □ *Line*

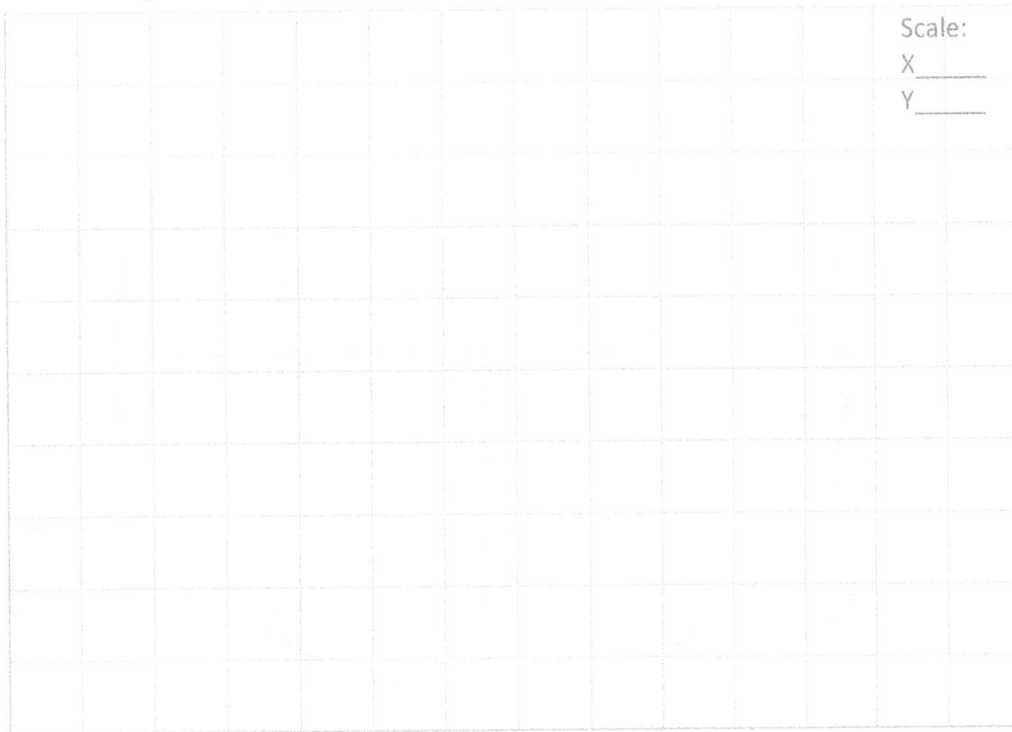

Scale:
X_____
Y_____

_____(___)

GRAPH SHOWING_____

B – Record in each box below the number of organisms in each environmental condition

TABLE 2 SHOWING THE NUMBER OF MILLIPEDES/ WOODLICE IN MOIST, DRY, LIGHT AND DARK CONDITIONS

Conditions	Number of millipedes/ woodlice			
	Count 1	Count 2	Count 3	Average
Moist, Light				
Dry, Light				
Moist, Dark				
Dry, Dark				

Note: Two variables are being manipulated at the same time here.

Additional observations: _____

_____

_____

Add a **_labelled drawing_** of the choice chamber to show the average preference/location of the organisms at the end of the experiment.

TITLE:_____

DISCUSSION

LIMITATIONS _____

_____

_____

_____

_____

PRECAUTIONS _____

_____

_____

SOURCES OF ERROR _____

_____

_____

_____

CONCLUSION *(What was the response of organisms to moisture and the response to light?)*

_____

_____

_____

_____

_____

REFLECTION – *What aspect of this experiment was most meaningful to you? How can this understanding be useful to you as a learner?*

_____

_____

_____

DISCUSSION QUESTIONS

1. Explain why the apparatus set up is called a choice chamber.
2. Explain how the behaviour of the organisms is advantageous for their survival.
3. Where in the chamber were most organisms expected to be found? Did the results show this? Refer/ quote values from the graph.

4. Identify what factors were kept constant for the results in Table 1.
5. How can the accuracy of the results be improved/ method modified?

6. Identify three suitable precautions when handling the organisms to ensure accuracy.
7. Are there any limitations in conducting this experiment? (Hint – organisms may be damaged, or had some disease/ virus that made their response atypical.)
8. What conclusions can be made from the results?

NOTES – IRRITABILITY AND TAXIS

Choice chambers are apparatus (dishes/ boxes) which contain sectioned areas of different conditions which act as environmental stimuli for investigating responses of organisms. Organisms, usually small invertebrates are placed inside the lid, and their choice for a particular condition is noted by counting and recording the numbers in each area after some time.

In general, woodlice prefer damp environments and dark environments. Therefore it is expected that more woodlice would be found in damp and dark sections of a choice chamber. **The movement of an animal in response to a stimulus is called "taxis".** **Photo-taxis** is movement due to light stimuli. **Kinesis** is the change in speed of movement in response to a stimuli. In general, woodlice tend to cluster in crevices and slow down, so they may have to be gently disturbed by tapping the chamber. Positive responses result in movement towards a stimulus, whereas negative responses result in movement away from a stimulus.

Woodlice are **isopods**, similar to crustaceans. They eat rotting plants and breathe through gill like structures and are thus adapted to moist habitats. They exhibit diurnal rhythms and prefer to come out mostly at night to avoid dry/ low humidity conditions during the day. In moist habitats water loss and drying out through the cuticle is prevented. Additionally, in dark areas, there is less chance of being spotted by predators, ensuring survival.

In this experiment the moisture content as well as light available were manipulated. However, there are a number of constants. The number of organisms used each time, the time left for acclimation before counting. The type of organisms should all be the same as well as the temperature of each environment. A suitable control would be the same choice chamber without any cotton or silica gel in the lower chambers.

Modifications to the method include – placing the same number of organisms in each of the different environments at the start of the experiment, and then after a suitable time counting the numbers in each area. More organisms can also be used instead of just 10 -12. Additional trials/ repeat experiments can also improve accuracy.

PAGE INTENTIONALLY LEFT BLANK

NAME:_____ CLASS:_____ **SKILL: PD - Draft**

24.   INVESTIGATIVE PROJECT - PART A – THE PROPOSAL (Draft document)

OBSERVATION

Some students noticed that some pumpkin seeds from the cafeteria waste in the school's compost heap germinated faster and grew bigger than some that fell on the sandy soil nearby.

Plan and design an experiment to test a suitable hypothesis on germination/ soil types.

Questions to consider – Developing the hypothesis:

_____

_____

_____

_____

_____

_____

_____

HYPOTHESIS _____

_____

_____

AIM _____

_____

APPARATUS AND MATERIALS

_____

_____

DIAGRAM

METHOD

Write the method as instructions.

The method must be a series of steps that flow logically. Use the checklist on the right to ensure the method is planned correctly.

☐   Main apparatus use
☐   Exact quantities of volume, mass, length, etc.
☐   How often measurements must be taken
☐   Time frame for entire experiment
☐   Repeats/ trials – for accuracy
☐   Values to collect as results/ measurements
☐   Control experiment set up identified
☐   How to interpret results.

_____

_____

_____

_____

_____

_____

_____

_____

_____

_____

_____

_____

_____

_____

VARIABLES

MANIPULATED:_____

RESPONDING: _____

CONSTANT: _____

_____

*Note: Constant variables are ALL the things kept the same throughout the experiment. There should be more than one constant variable. However, there is only 1 manipulated variable and 1 responding variable. **The Control experiment set up IS NOT the constant variable!***

EXPECTED RESULTS *(In the space available, write a description of results expected AND draw up a blank table/ graph to collect your results)*

EXPLANATION OF EXPECTED RESULTS *(Relate the known theory to the expected results)*

_____

_____

_____

_____

_____

_____

_____

STATEMENT OF ACCEPTANCE/REJECTION OF HYPOTHESIS

_____

_____

_____

_____

LIMITATIONS

_____

_____

_____

_____

ASSUMPTIONS

_____

_____

_____

_____

REFLECTION

_____

_____

_____

The criteria marked when planning and designing an experiment should have all the sections outlined.

CRITERIA	EXPLANATION OF CRITERIA	MARKS
HYPOTHESIS	Clearly stated (plausible) Testable	2
AIM	Related to hypothesis (indicates the manipulated and responding variable; begins with "to")	1
MATERIALS & APPARATUS	Appropriate – use of lab apparatus List – all that will be used	1
PROCEDURE/ METHOD	INSTRUCTUIONAL tense; will allow achievement of aim – LOGICAL STEPS (1) Repetition of steps for accuracy (1) A suitable control is set up (1)	3
VARIABLES	1. manipulated – 1 variable changed 2. responding – 1 variable recorded 3. constant (NOT controlled) – many variables kept the same	
EXPECTED RESULTS	Reasonable – draw a table, with headings and title if you have to collect results; draw a graph/ Link with method	1
EXPLANATION OF EXPECTED RESULTS	Describe in prose (paragraphs) what you expect to observe and why. Identify what results will make you accept the hypothesis and why. Identify what results will make you reject the hypothesis and why.	1
ASSUMPTIONS/ LIMITATIONS/ PRECAUTIONS/ POSSIBLE SOURCES OF ERROR Explain any one of the above		1

NAME:_____CLASS:_____ SKILL: **PD - Final**

25. INVESTIGATIVE PROJECT - PART A – THE PROPOSAL: FINAL

OBSERVATION:

Some students noticed that some pumpkin seeds from the cafeteria waste in the school's compost heap germinated faster and grew bigger than some that fell on the sandy soil nearby.

Plan and design an experiment to test a suitable hypothesis on germination/ soil types.

HYPOTHESIS _____

_____

_____

AIM _____

_____

APPARATUS AND MATERIALS

_____

_____

DIAGRAM

METHOD

Write the method as instructions.

□   Main apparatus use
□   Exact quantities of volume, mass, length, etc.
□   How often measurements must be taken
□   Time frame for entire experiment
□   Repeats/ trials – for accuracy
□   Values to collect as results/ measurements
□   Control experiment set up identified
□   How to interpret results.

_____

_____

_____

_____

_____

_____

_____

_____

_____

_____

_____

_____

_____

_____

_____

_____

VARIABLES:

MANIPULATED:_____

RESPONDING: _____

CONSTANT: _____

_____

EXPECTED RESULTS

(In the space available, write a description of results expected AND draw up a blank table/ graph to collect your results)

_____

_____

_____

_____

_____

EXPLANATION OF EXPECTED RESULTS

(Relate the known theory to your expected results; add definitions and equations where necessary)

_____

_____

_____

_____

_____

_____

_____

_____

_____

_____

_____

STATEMENT OF ACCEPTANCE/REJECTION OF HYPOTHESIS

_____

_____

_____

LIMITATIONS_____

_____

_____

_____

ASSUMPTIONS_____

_____

_____

_____

REFLECTION_____

_____

_____

_____

_____

CRITERIA	EXPLANATION OF CRITERIA	MAX. MARKS	TEACHER MARKS
HYPOTHESIS	Clearly stated (plausible) Testable	2	
AIM	Related to hypothesis (indicates the manipulated and responding variable; begins with "to")	1	
MATERIALS & APPARATUS	Appropriate – use of lab apparatus List – all that will be used	1	
PROCEDURE/ METHOD	INSTRUCTUIONAL tense; will allow achievement of aim – LOGICAL STEPS (1) Repetition of steps for accuracy (1) A suitable control is set up (1)	3	
VARIABLES	1. manipulated – 1 variable changed 2. responding – 1 variable recorded 3. constant (NOT controlled) – many variables kept the same		
EXPECTED RESULTS	Reasonable – draw a table, with headings and title if you have to collect results; draw a graph/ Link with method	1	
EXPLANATION OF EXPECTED RESULTS	Describe in prose (paragraphs) what you expect to observe and why. Identify what results will make you accept the hypothesis and why. Identify what results will make you reject the hypothesis and why.	1	
ASSUMPTIONS/ LIMITATIONS/ PRECAUTIONS/ POSSIBLE SOURCES OF ERROR Explain any one of the above		1	

NAME:_____**CLASS:**_____ **SKILL: AI**

26.   INVESTIGATIVE PROJECT – PART B – IMPLEMENTATION

INTRODUCTION: *(Background to the problem; linked to the plan and design)*

_____

_____

_____

_____

_____

_____

_____

_____

_____

_____

_____

METHOD:

*Linked to Part A – the Proposal. (There should be a change of tense from instructions to **past tense**.)*

_____

_____

_____

_____

_____

_____

_____

_____

_____

_____

_____

_____

_____

_____

_____

_____

_____

_____

_____

_____

_____

_____

_____

RESULTS:

Arrange your worksheet below to collect results – this can be a table, a graph AND pictures. Insert additional pages as needed.

DISCUSSION

LIMITATIONS _____

_____

_____

PRECAUTIONS _____

_____

_____

SOURCES OF ERROR _____

_____

_____

_____

CONCLUSION _____

_____

_____

_____

REFLECTION _____

_____

_____

_____

_____

_____

_____

_____

_____

_____

_____

MARK SCHEME FOR IMPLEMENTATION OF INVESTIGATIVE PROJECT – AI SKILL

SECTION	BREAK DOWN OF MARKS	Max Mark	Teacher's Mark
METHOD	Linked to Proposal with a change of tense from instructions to **past tense**	1	
RESULTS	**Correct formulae and equations** Accurate (2); Acceptable (1)	4	
	Accuracy of data: Accurate (2); Acceptable (1)		
DISCUSSION	**Explanation** Development of points: Thorough (2); Partial(1)	5	
	Interpretation Fully supported by data (2); Partially supported by data (1)		
	Trends Stated (1)		
LIMITATIONS/ PRECAUTIONS/ SOURCES OF ERROR	-Sources of error identified (1) -Precautions stated (1) -Limitation stated (1)	3	
REFLECTIONS	- Relevance of the experiment to real life (Self, Society AND The environment) (1) - Impact of knowledge gain from experiment on self (1) - Justification for any adjustment made during experiment (1)	3	
COMMUNICATION of information	Use of appropriate scientific language, grammar and clarity of expression: - all of the time (2) - some of the time (1)	2	
CONCLUSION	- Stated (1) - Related to the aim (1)	2	
	TOTAL MARKS AWARDED	20	

27.   PLAN AND DESIGN # 2 – GLYCEMIC INDEX OF BANANAS

OBSERVATION

John is a healthy middle aged man. His family has a history of diabetes. He was recently told by the doctor to monitor his sugar intake and eat food with a low glycemic index (GI) as a preventative measure to avoid becoming diabetic.

On doing some research he found that the GI number is a ranking of carbohydrates in foods from 1 to 100, according to how they affect the blood glucose a few hours after eating. Foods with a low GI of 55 or less are slowly metabolised causing a slower rise in blood glucose. Low GI foods are good for diabetics and pre-diabetics as they usually do not produce sufficient insulin. High GI foods are useful also as they help with energy for recovery after exercise.

John likes bananas and noticed that the fruits became softer, moister and sweeter when ripen. He wondered if the state of ripeness of the banana had an effect on the GI, and which stage of ripeness was best for him to eat.

Plan and design an experiment based on John's observations. Assume that you only have access to high school lab resources and bananas at different stages of ripeness.

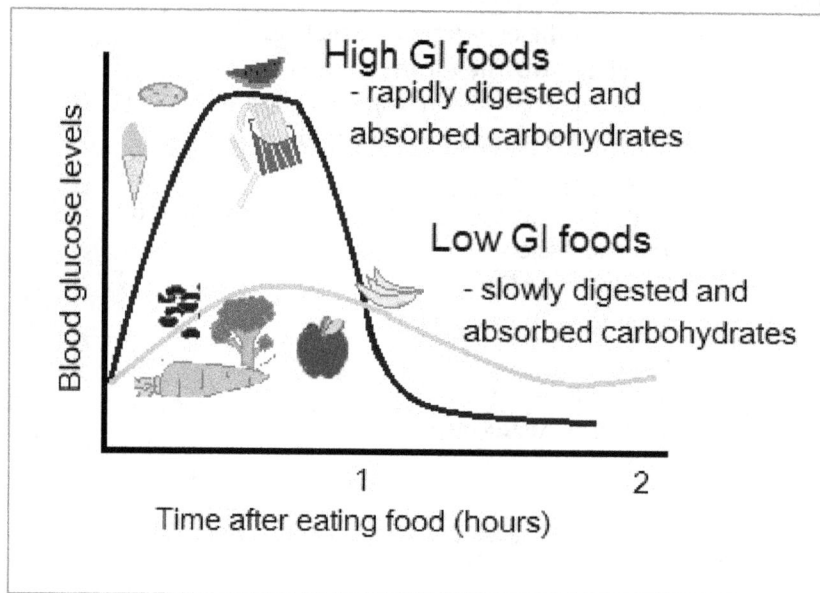

SEE PAGE 230 FOR ADDITIONAL SCENARIOS THAT MAY BE USED TO PLAN AND DESIGN AN EXPERIMENT.

NAME:_____ CLASS:_____ SKILL: PD

OBSERVATION FOR PD # 2

_____

_____

_____

_____

_____

_____

_____

HYPOTHESIS

_____

_____

AIM

_____

_____

APPARATUS AND MATERIALS

_____

_____

DIAGRAM:

METHOD

Write the method as instructions. *Use the check list on the right to ensure the method is complete.*

☐   Main apparatus use
☐   Exact quantities of volume, mass, length, etc.
☐   How often measurements must be taken
☐   Time frame for entire experiment
☐   Repeats/ trials – for accuracy
☐   Values to collect as results/ measurements
☐   Control experiment set up identified
☐   How to interpret results.

_____

_____

_____

_____

_____

_____

_____

_____

_____

_____

_____

_____

_____

_____

_____

_____

_____

VARIABLES

MANIPULATED:_____

RESPONDING: _____

CONSTANT: _____

_____

_____

EXPECTED RESULTS

(In the space available, write a description of results expected AND draw up a blank table to collect your results)

_____

_____

_____

_____

_____

EXPLANATION OF EXPECTED RESULTS

(Relate the known theory to your expected results; add definitions and equations where necessary)

_____

_____

_____

_____

_____

_____

_____

_____

_____

_____

_____

STATEMENT OF ACCEPTANCE/REJECTION OF HYPOTHESIS

_____

_____

_____

LIMITATIONS

_____

_____

_____

ASSUMPTIONS

_____

_____

_____

REFLECTION

_____

_____

_____

_____

CRITERIA	EXPLANATION OF CRITERIA	MAX. MARKS	TEACHER MARKS
HYPOTHESIS	Clearly stated (plausible) Testable	2	
AIM	Related to hypothesis (indicates the manipulated and responding variable; begins with "to")	1	
MATERIALS & APPARATUS	Appropriate – use of lab apparatus List – all that will be used	1	
PROCEDURE/ METHOD	INSTRUCTUIONAL tense; will allow achievement of aim – LOGICAL STEPS (1) Repetition of steps for accuracy (1) A suitable control is set up (1)	3	
VARIABLES	1. manipulated – 1 variable changed 2. responding – 1 variable recorded 3. constant (NOT controlled) – many variables kept the same		
EXPECTED RESULTS	Reasonable – draw a table, with headings and title if you have to collect results; draw a graph/ Link with method	1	
EXPLANATION OF EXPECTED RESULTS	Describe in prose (paragraphs) what you expect to observe and why. Identify what results will make you accept the hypothesis and why. Identify what results will make you reject the hypothesis and why.	1	
ASSUMPTIONS/ LIMITATIONS/ PRECAUTIONS/ POSSIBLE SOURCES OF ERROR Explain any one of the above		1	

NAME:_____ CLASS:_____ **SKILL: AI_____**

TOPIC: GENETICS   SYLLABUS OBJECTIVE: C 2.10

28.   INVESTIGATION TO DETERMINE THE SEX OF AN OFFSPRING IN HUMANS

AIM:_____

_____

Who or what determines if a baby is a boy or a girl? The mother or the father? It's the father! And the resulting combination of sex chromosomes in his sperm. The sex chromosome provided by the father/male parent contains either an X or a Y chromosome which ultimately determines the gender. In this experiment you will use coloured paper squares to investigate exactly how the possible combination of sex chromosomes occurs. You will also understand that sex determination is based on the probability of obtaining a particular sex chromosome in the male gamete, and this is random/due to chance.

APPARATUS AND MATERIALS:

- 2 Beakers
- 2 Petri dishes
- Labels
- Squares of blue paper – 50

- Squares of pink paper – 75
- Students
- Blindfold (optional)

DIAGRAM:

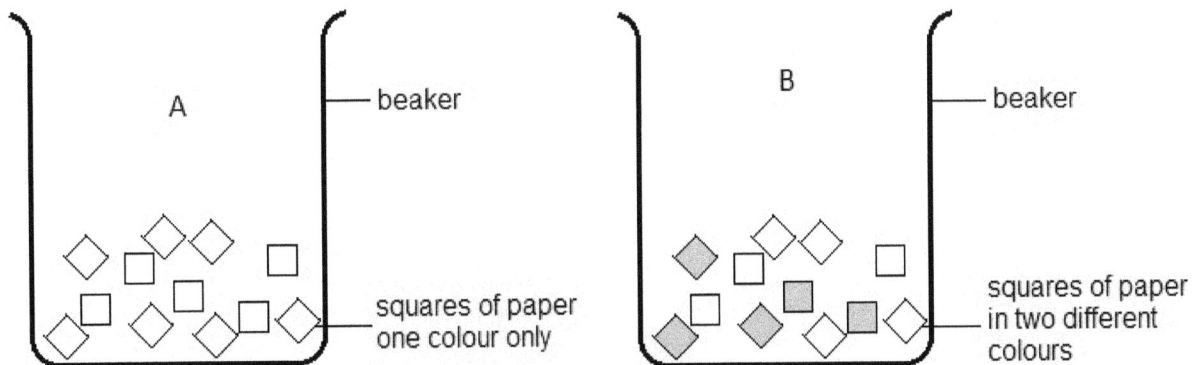

DRAWING SHOWING SET-UP OF BEAKERS – A – SAME COLOUR SQUARES; B – TWO COLOUR SQUARES

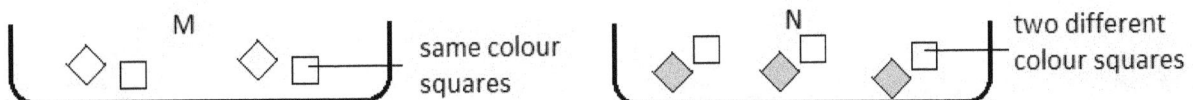

DRAWING SHOWING PETRI-DISHES TO PLACE EITHER M – THE SAME COLOUR SQUARES; N – DIFFERENT COLOUR SQUARES

INSTRUCTIONS:

1. Label 2 beakers A and B respectively. Label 2 Petri Dishes M and N.
2. Draw up a table as shown in the results with 3 columns headers (trial, M and N)
3. Count and place 50 pink squares in a beaker labelled A.
4. Count and add 25 pink squares and 25 blue squares to beaker labelled B.
5. Thoroughly mix the squares in beaker B.
6. Place the beakers side by side, and have a student close his/her eyes and choose one paper square from each beaker.
7. If the papers are both the same put them in Petri dish M. Record a tick under the column for M in your table.
8. If the papers are different (one blue, one pink) then place them in Petri dish N, and record a tick for that trial.
9. Repeat the selection process for a total of 10 trials. Each trial should have only 1 tick either for M or for N.
10. Total the number of ticks for M and for N at the end of the 10 trials and state this as the observed ratio of M to N.

Rewrite your method in past tense in the space below or on a separate page.

METHOD

_____

_____

_____

_____

_____

_____

_____

_____

_____

_____

_____

_____

_____

_____

RESULTS:

TABLE SHOWING: _____

Trials	M – same colour squares	N – different colour squares
1		
2		
3		
4		
5		
6		
7		
8		
9		
10		
Total ticks recorded		
Genotype represented		
Gender (phenotype)		
Observed Ratio		

Place a tick ✓ under the appropriate column for each trial/ combination of squares.

DISCUSSION

LIMITATIONS _____

_____

_____

_____

_____

_____

PRECAUTIONS _____

_____

_____

_____

_____

SOURCES OF ERROR _____

_____

_____

_____

_____

CONCLUSION _(How is sex/ gender determined in humans?)_

_____

_____

_____

_____

REFLECTION _(What do you have a better understanding/ appreciation for with respect to inheritance of sex chromosomes in man?)_

_____

_____

_____

_____

_____

DISCUSSION QUESTIONS: *(Answer ALL)*

1. With respect to inheritance, explain the purpose of:
    a. The blue squares and the pink squares.
    b. The beaker.
    c. The petri dish.
2. Why were the experimenter's eyes closed when taking paper squares?
3. Draw a genetic diagram to show the inheritance of the sex chromosomes from a male and female parent. State the expected ratio of boys to girls in the offspring.
4. Compare the expected ratio (from the genetic diagram in 3.) with the observed ratio of 'boys' to 'girls' from the results. Are they the same? Why?

Genetic Diagram Steps:
Parental phenotypes: _____ x _____
Parental genotypes: _____ x _____
Gametes : ⃝⃝ x ⃝⃝
Random fertilization:
Offspring genotypes:
Offspring phenotypes:
Offspring phenotypic ratio:

5. In what ways does the experiment differ from what really happens? (Time, number of children, number of gametes made by male versus female.)
6. Identify at least two precautions when making selections. Explain any sources of error.
7. Identify the limitations of this experiment. (Choice of material for gametes/ real life)

NOTES – INHERITANCE OF SEX CHROMOSOMES IN MAN – GENDER DETERMINATION

What determines if a baby is a boy or a girl? Inheritance of the sex chromosomes!

-   Females have two X (sex) chromosomes (genotype XX),
-   Males have one Y chromosome and one X chromosome (genotype XY).
-   So if a baby inherits an **X from the father** and an X from the mother, **it will be a girl.**
-   If the baby inherits **a Y chromosome from the father** and an X from the mother, **it will be a boy**.

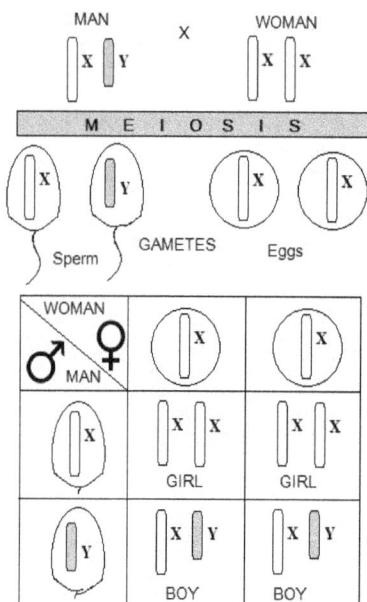

Notice that **a mother can only pass on an X chromosome** (100% chance), so **the sex of the baby is determined by the father's contribution**. The father has a 50 % chance of passing on his Y chromosome and a 50% chance of passing on his X chromosome. Hence there is a 50 percent chance that a child will be a boy (XY) or a girl (XX) creating an expected ratio of 1:1 boys to girls. Still why does the observed ratio sometimes differ from the expected?

Inheritance of sex follow the same laws of probability that apply when tossing a coin.

- There are two sides, so there is a 50% chance of getting heads and a 50% chance of tails – this applies to the X or Y chromosome in the male gamete.
- Additionally the outcome of 1 coin toss has no impact on the outcome of the next toss. This also applies to the sex of offspring each time a couple has a child.
- Furthermore, the larger the sample size (more tosses of a coin), the closer to the expected/ predicted probability/ ratios. Therefore 10 trials may not be enough chances, and 30 trials may be better.

NAME:_____ CLASS:_____ SKILL: ORR____

TOPIC: GENETICS   Syllabus Objective: C2.8, 2.9, 3.1

29.   INVESTIGATING INHERITANCE OF TRAITS USING COINS

AIM: TO INVESTIGATE HOW GENES FROM THE MOTHER AND FATHER DETERMINE A CHILD'S APPEARANCE

Do you ever notice that your brothers and sisters have similar features to you? Do you have some facial features from your mom and some from your dad? How does this occur? Do you have a feature that neither your mother nor your father has? How is this even possible? In this experiment, you will see how genes from the mother and father determines their children's appearances.

APPARATUS AND MATERIALS:

- Brown paper – 5 ovals (15cm x 10cm)/ cut up 6lb brown paper bags
- Facial features outlines (make 3 copies of this page)
- Scissors
- Glue or tape
- Two different coins – 25cent piece and 5cent piece OR red and black checkers
- Markers/ colouring pencils – optional.
- Round circular dot labels with the numbers 1 and 2 (to stick over the sides of the coin/checkers)

DIAGRAM:

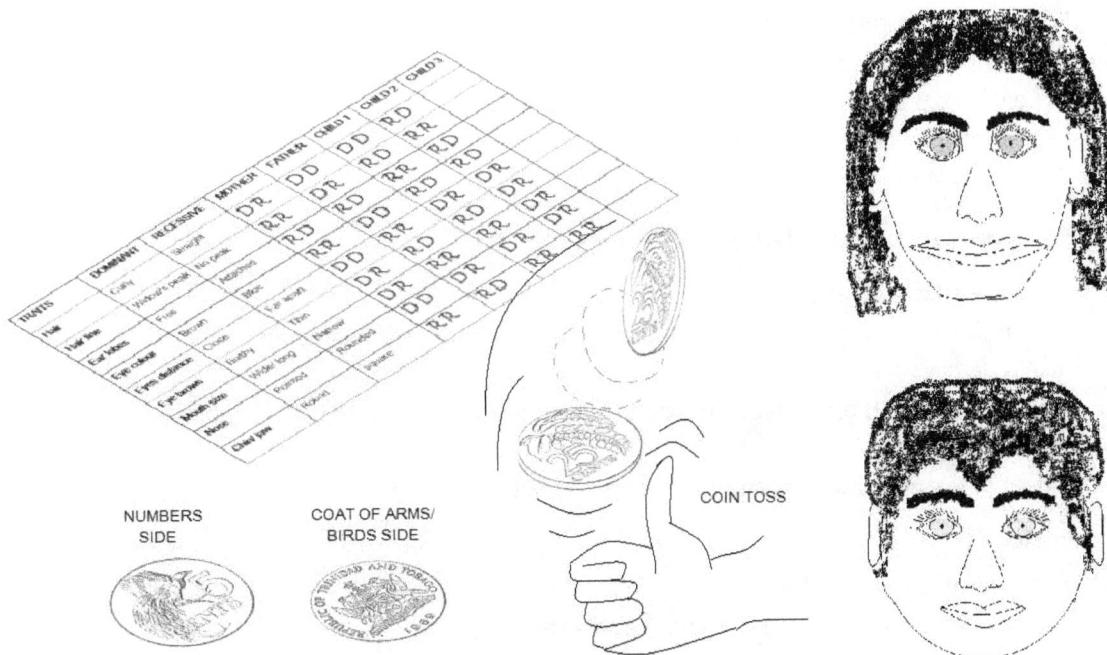

DIAGRAM SHOWING HOW THE COIN TOSS IS DONE TO SELECT PARENTAL GENES AND CREATE FACES

METHOD:

A – Creating the parents

1. Designate the 25 cent piece as the mother and the 5cent piece as the father.
2. For each coin, the numbers side will be D (for dominant allele) and the coat of arms/birds side will be R (for recessive allele)
3. Generate the genetic profile of the mother by flipping the 25 cent piece **twice**. Each time, record D or R under the mother alleles column for that trait in the results.
4. Repeat step 3 for all the traits of the mother. Write out the corresponding phenotypes.
5. Generate the genetic profile of the father by flipping the 5 cent piece **twice**. Each time, record D or R under the father alleles column for that trait. Repeating until all traits and phenotypes are determined.
6. Label one brown paper oval MOTHER and another FATHER.
7. Using the facial features outlines, cut out and stick the features determined for each parent and create the mother and father on the brown paper ovals.

B – Creating the children

1. Decide firstly if the child will be a boy or a girl. Flip the 5 cent piece and if it is numbers let it be a boy; if it is coat of arms/birds, let it be a girl. Label an oval child 1 "boy" or "girl" as appropriate.
2. Now the sides of the coin will be used differently – Label the numbers side 1st to represent the first allele from the parent and label the coat of arms/birds 2nd to represent the second allele obtained from the parent.
3. Starting with the hair trait, flip both coins at the same time. For the 25c record if the first or second allele obtained by the mother will be used for this child's hair allele. Do the same for the 5c piece to determine the father's allelic contribution.
4. Determine the phenotype for the hair trait.
5. Repeat step 3 and 4 for each of the traits of child 1.
6. Repeat the entire process – steps 1 to 5 to determine the traits of child 2 and child 3.
7. Compare your family with another group's family.

C – Using a punnet square to determine phenotype.

1. Choose 1 trait from the table 1.
2. Define the alleles using a capital letter for the dominant trait allele and a common letter for the recessive trait allele.
3. Using the parental phenotypes for that trait, determine the possible combinations of the offspring.
4. Did any of your 3 children above show these traits?

ASSUMPTIONS:
- All traits are controlled by 1 gene only
- Each gene has only 2 alleles – a dominant and a recessive.
- There is complete dominance, and no other gene interaction like co-dominance or incomplete dominance.

Note, in reaility, some of these traits are controlled by more than one gene (polygenetic traits).

RESULTS:

TABLE 1. GENETIC TRAITS AND COMBINATION OF ALLELES USED TO DETERMINE PARENTAL PHENOTYPES

TRAITS	DOMINANT - D (Numbers)	RECESSIVE - R (Coat of arms)	MOTHER			FATHER		
			1st, toss allele	2nd toss allele	Phenotype	1st, toss allele	2nd toss allele	Phenotype
HAIR	Curly	Straight						
HAIR LINE	Widow's peak	No widow's peak						
EAR LOBES	Free	Attached						
EYE COLOUR	Brown	Blue						
EYES DISTANCE	Close	Far apart						
EYE BROWS	Bushy	Thin						
MOUTH SIZE	Wide/ long	Narrow						
NOSE	Pointed	Rounded						
CHIN/ JAW	Round	square						

In this coin toss: **Numbers** – represents dominant allele   **Coat of arms** – represents recessive allele

- Recall the dominant phenotype is obtained if (i) there are two dominant alleles – DD or (ii) if there is a dominant and a recessive allele – either DR or RD.
- Only when there are two recessive alleles – RR, will the recessive phenotype be obtained.

TABLE 2. COMBINATIONS OF ALLELES USED TO DETERMINE CHILDRENS' PHENOTYPES

TRAITS	Child 1 _____			Child 2 _____			Child 3 _____		
	Alleles		Phenotype	Alleles		Phenotype	Alleles		Phenotype
	M	F		M	F		M	F	
HAIR									
HAIR LINE									
EAR LOBES									
EYE COLOUR									
EYES DISTANCE									
EYE BROWS									
MOUTH SIZE									
NOSE									
CHIN/ JAW									

M = mother's allele from table 1   F = father's allele from table 1

In this coin toss: **Numbers 1st** – represents the first allele obtained by the parent in table 1

Coat of arms 2nd – represents the second allele obtained by the parent in table 1

Stick a copy of the mother and the father in the spaces below.

MOTHER

FATHER

GENETIC CROSS TO SHOW THE INHERITANCE OF ONE TRAIT

Genetic trait selected for the cross: _____

Defining the alleles:Let _____ be the allele to represent _____ (Dominant)

Let _____ be the allele to represent _____ (Recessive)

Genetic Diagram Steps:

Parents: Female   xMale

Parental phenotypes: _____   x   _____

Parental genotypes: _____   x   _____

Gametes :   ◯ ◯ x   ◯ ◯

Random fertilization:

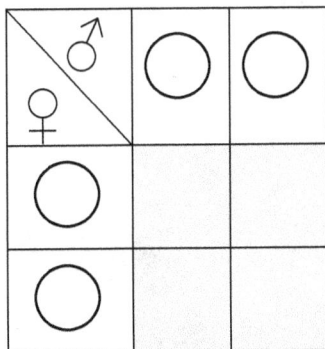

	♂ ◯	◯
♀ ◯		
◯		

Offspring genotypes:_____

Offspring phenotypes:_____

Offspring phenotypic ratio: _____

FACIAL FEATURES OUTLINES

CURLY HAIR - FEMALE

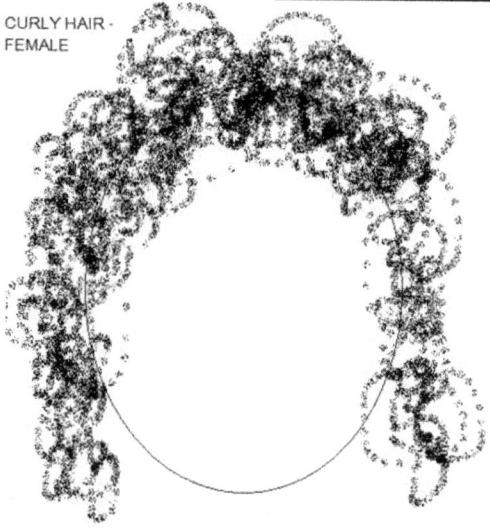

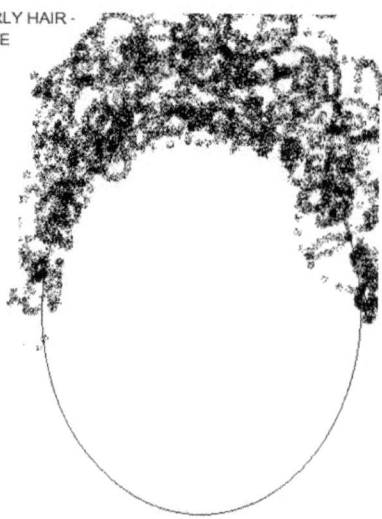

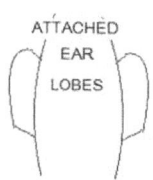

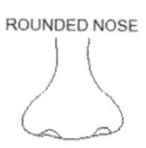

CURLY HAIR - MALE

FREE EAR LOBES

ATTACHED EAR LOBES

POINTED NOSE

ROUNDED NOSE

WIDOW'S PEAK FOR CURLY HAIR

WIDOW'S PEAK FOR STRAIGHT HAIR

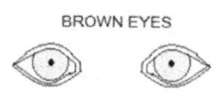

BROWN EYES

BLUE EYES

STRAIGHT HAIR - FEMALE

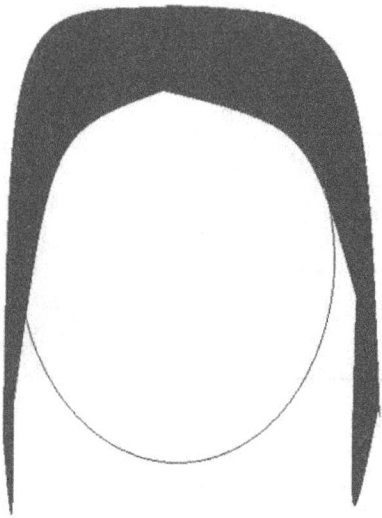

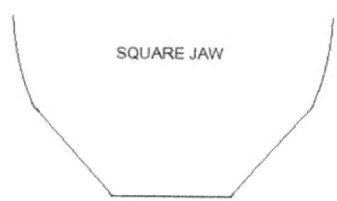

STRAIGHT HAIR - MALE

NARROW MOUTH

WIDE/ LARGE MOUTH

BUSHY EYEBROWS

THIN EYEBROWS

SQUARE JAW

DISCUSSION_____

_____

_____

_____

_____

_____

_____

_____

_____

_____

_____

_____

_____

_____

_____

_____

_____

_____

LIMITATIONS_____

_____

_____

PRECAUTIONS_____

_____

SOURCES OF ERROR_____

_____

CONCLUSION – *What was found out in this lab about inheritance of traits?*

_____

_____

REFLECTION – *How are you impacted? What do you have a better understanding or appreciation of?*

_____

_____

_____

DISCUSSION QUESTIONS:

1. Define the following terms – gene, allele, dominant, recessive, homozygous and heterozygous.
2. Which biological processes does the coin toss model?
3. Do all of the children look alike or are there some differences among them? Why?
4. Do the children resemble the parents in any way? How?
5. Is there any feature where a child had a different phenotype from either of his/her parents? How is this possible (Hint two recessive alleles can come together from heterozygous parents).
6. What does the genetic diagram/ punnet square show? How is this different from tossing the coins to determine the children's genotype?
7. Are all traits that exist naturally in humans controlled by one gene only?
8. Identify any precautions taken to obtain fair results. Were there any sources of error?

NOTES – INHERITANCE OF GENETIC TRAITS

- A **gene** is a segment of DNA that gives instructions for making a protein.
- There are different versions of the same genes called **alleles**.
- Different alleles can make different proteins which result in different observable characteristics.
- The cells of the body contains two copies of each gene, or two alleles.
  - One is inherited from your mother.
  - The other is inherited from your father.
- If both copies of the gene have the same allele, the person has a **homozygous genotype**.
- If the two copies of alleles are different, the person has a **heterozygous genotype**.
- **Dominant alleles**, when present are expressed, even if the person is heterozygous.
- **Recessive alleles** are masked by the dominant allele in the heterozygous condition; but are expressed in the homozygous condition.
- Sometimes recessive alleles create defective proteins, but not always.

A two-sided coin is used to model the way genes behave during meiosis and fertilization. For the parents, numbers represented the dominant allele and coat of arms/birds represented the recessive allele. Tossing the coin represents that there is equal probability of getting a dominant or a recessive allele, as is the case when gametes are made during meiosis. When the two coins are tossed together for the children, the result of this pair indicated the genotype of the child which is similar to what occurs during fertilization and formation of the zygote.

Each child that is created using the alleles of the parents results in some similarities and/or differences to each other and to either parent. Similar features between siblings and parents will be due to the dominant alleles inherited. When the children differ from their parents, this is due to a homozygous recessive genotype resulting for that trait (a recessive allele from each parent combined).

The punnet square is a simplified way of determining all the possible combinations of offspring genotypes, based on the parents' genotypes. It is a model that takes into account all the possibilities. In this experiment, for the punnet square, only one trait was considered at a time – monohybrid inheritance. However, when tossing the coins only 3 times (for 3 children) not all possibilities may be observed. Therefore punnet square predictions are accurate for large samples but not for small samples such as individual families.

NAME:_____CLASS:_____ SKILL:__ORR/AI____

TOPIC: VARIATION   SYLLABUS OBJECTIVE: C 3.1, 3.2, 3.3

30.   <u>INVESTIGATING THE VARIATION IN HEIGHTS AND OTHER TRAITS IN A CLASS</u>

AIM:_____

Are your genes the only factor which determine the way you look? Can the environment influence the type of ear lobes you have or your ability to roll your tongues, what about your size/ heights? All individuals in a population/class have certain characteristics (phenotypes) that make them different from one another. Whereas some of these variations are spread over a range of measurements, some variations are either present or absent. Genes and the environment may or may not influence some of these characteristics which show variation. In this experiment you should be able to distinguish between continuous and discontinuous variation among the class of students.

APPARATUS AND MATERIALS:
- Class of students
- Height chart/ fixed meter ruler
- Masking tape/ double sided tape
- Long ruler (30cm)
- Data capture sheet.

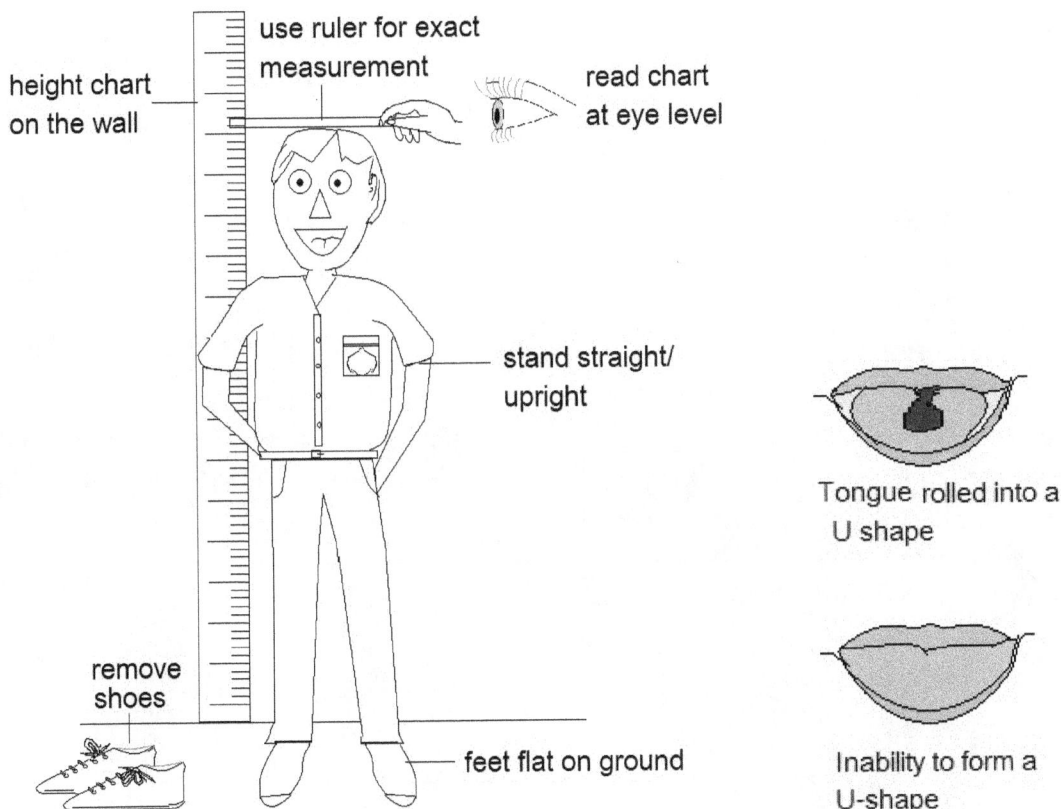

<u>DRAWING SHOWING HOW TO MEASURE HEIGHT OF STUDENTS AND SHAPE OF TONGUE WHEN ROLLED.</u>

INSTRUCTIONS:

1. Obtain an alphabetical class listing.
2. Work in pairs to measure and record the height of your partners.
3. The person being measured should remove his/her shoes and stand upright.
4. The second person should read the exact height in cm to the nearest 0.1cm, with the aid of the smaller ruler placed perpendicular to the height chart on the wall.
5. Pool the class heights data on one sheet, and share with each other.
6. Convert the raw data for heights by grouping heights into a frequency data table and then tallying the number of individuals in each height group.
7. Plot a histogram graph to show the distribution of heights in the class.
8. Add a line to connect the midpoints of each bar of height grouping.
9. Observe another characteristic that the students have – ability/inability to roll tongues.
10. Tally the class data and plot a bar graph to show the number of students that have the ability to roll their tongues versus those that cannot roll their tongues.

Rewrite your method in past tense in the space below or on a separate page.

METHOD:

Date:_____ Lab #:_____ Page No:_____

RESULTS: *(Fill in each of the tables and graphs below, adding the necessary TITLES and data.)*

TABLE 1 SHOWING: _____

Name of Student	Height (cm)	Ability to Tongue Roll	Name of Student	Height (cm)	Ability to Tongue Roll

TABLE 2 SHOWING _____

Height Group (cm)	Tally	Number of Students (frequency)	Midpoint of height group (x)	Midpoint x frequency f (x)
130.0 – 134.9				
135.0 – 139.9				
140.0 – 144.9				
145.0 – 149.9				
150.0 – 154.9				
155.0 – 159.9				
160.0 – 164.9				
165.0 – 169.9				
170.0 – 174.9				
175.0 – 179.9				
TOTAL	-			

Graph 1 – Draw a histogram to show the distribution of heights for a class of students.

□ *Title*   □ *Scale*   □ *Labels on both axes*   □ *Units in each label*   □ *Plots correct*   □ *Line*

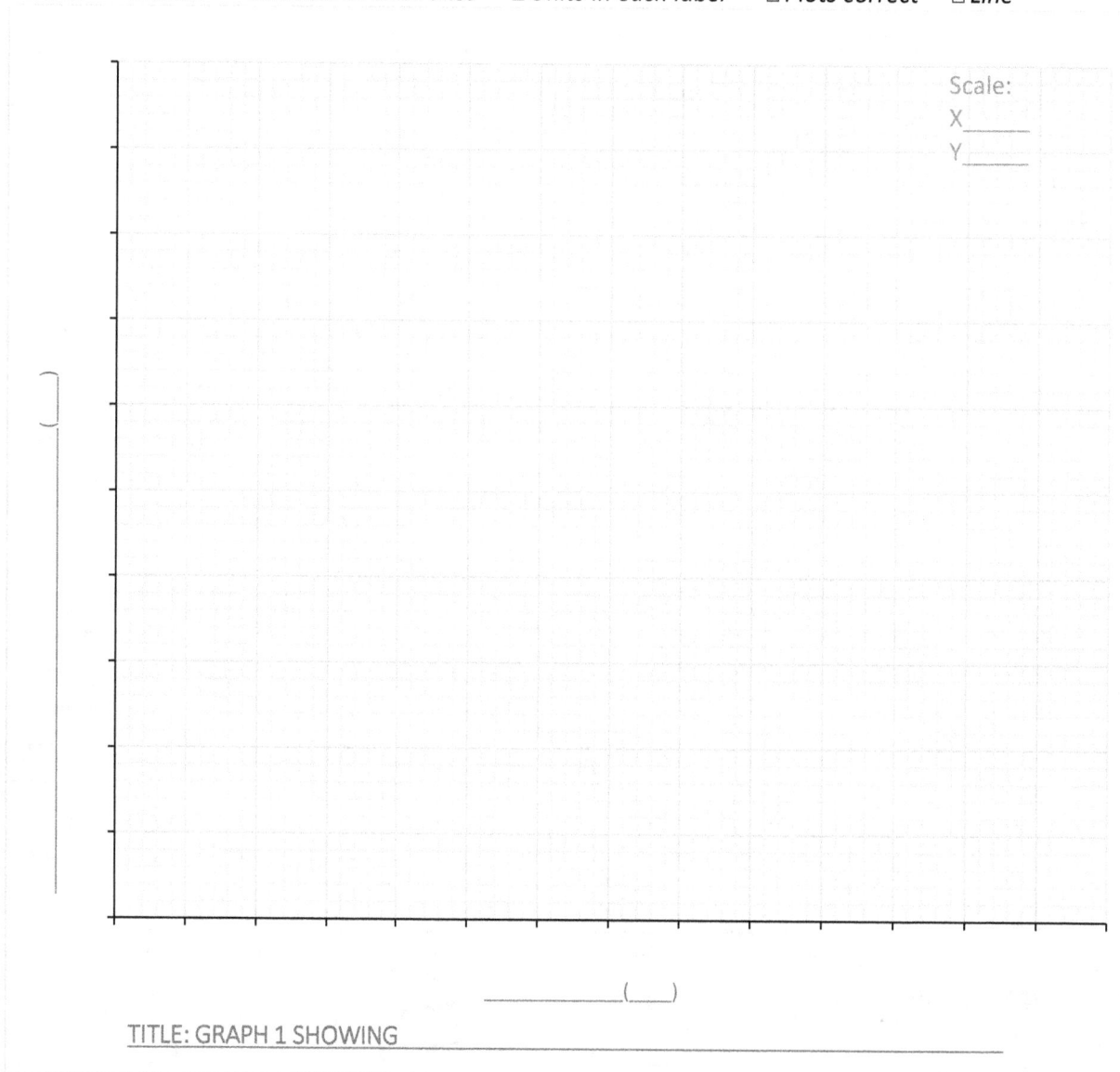

Scale:
X_____
Y_____

_____(____)

TITLE: GRAPH 1 SHOWING _____

Statistics from Graph 1 – Histogram of height for a class of students:

- Height ranges (shortest to tallest): _____
- Estimated Mean height: $\Sigma fx / \Sigma f$ where (Σ means "the sum of")

  _____

- Estimated Median height :_____
- Estimated Mode height: (frequency with most students) _____

TABLE 3 SHOWING _____

Ability to Roll tongues	Number of students
Able to roll tongues – tongue rollers	
Not able to roll tongues – non-rollers	
TOTAL	

Graph 2 – Bar chart to show the numbers of students that are able to roll their tongues.

□ Title   □ Scale   □ Labels on both axes   □ Units in each label   □ Plots correct   □ Line

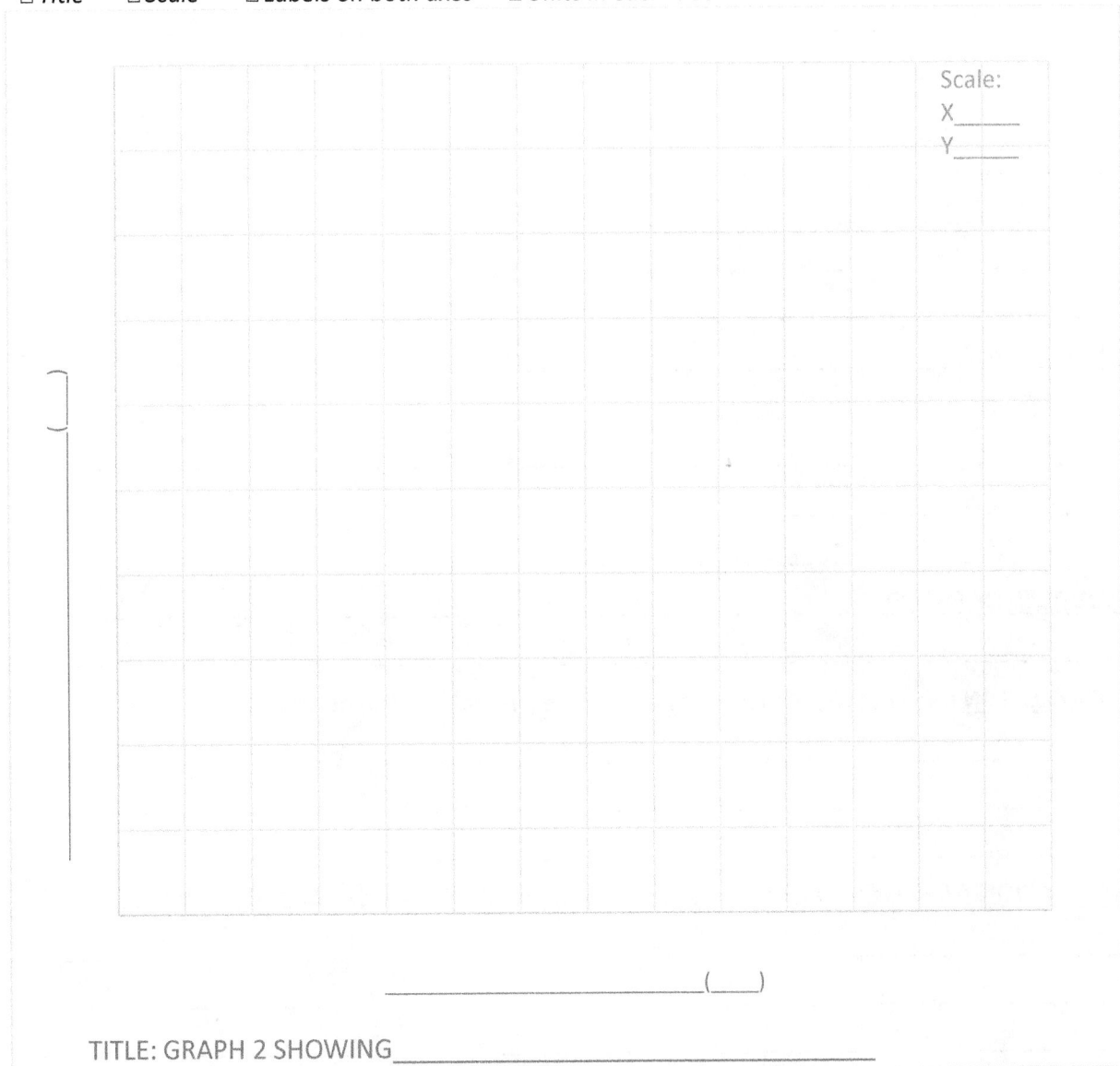

Scale:
X_____
Y_____

_____(____)

TITLE: GRAPH 2 SHOWING_____

DISCUSSION

_____

_____

_____

_____

_____

_____

_____

_____

_____

_____

_____

_____

_____

_____

_____

_____

LIMITATIONS

_____

_____

PRECAUTIONS

_____

SOURCES OF ERROR

_____

CONCLUSION (How does height vary? There are more....(tongue rolling ability) than....)

_____

_____

REFLECTION (How is this relevant to your everyday life? What do you know now that....)

_____

_____

_____

DISCUSSION QUESTIONS:

1. What two major factors influence phenotypic variation among a population?
2. Why is variation important?
3. Why were different types of graphs used for each trait (continuous/discontinuous)?
4. Based on the results,
   a. Describe the overall shape of the histogram (most persons were...., least were ...)
   b. Explain why this result was expected/not expected for heights compared to the theory (usually heights follow a normal distribution curve)?
   c. Describe the bar chart for tongue rolling ability. Was this expected? Explain.
   d. Explain sources of error that may have made any results inaccurate.
5. Identify precautions taken when measuring heights.
6. Explain any two limitations of this experiment (all of the students are not the same age, therefore, there is a chance that could influence heights; the reasons for variation in height have not been identified.)

VARIATION NOTES "Phenotype = Genotype + Environment"

Within a species, there is usually a great deal of phenotypic variation between individuals. Variation is important as it is the raw material for natural selection allowing for adaptations and extinction is less likely. Variations may be A) inherited or B) acquired (due to the environment)

(A) **Inherited variations** result from the **activity of genes**
   - They are genetically controlled,
   - They cannot be altered by the environment,
   - For example, blood groups, finger prints and gender (sex); these traits cannot be changed naturally

(B) **Acquired characteristics** also result from gene expression but are influenced by **environmental conditions** during a lifetime such as activities or nutrition:
   - Examples of acquired conditions in humans: language, obesity, athletic skills, mental skills, body building, sun tan
   - Acquired characteristics **cannot be inherited**

GENETIC CAUSES OF VARIATION

• Meiosis
  – Crossing over
  – Random assortment
• Sexual reproduction
• Mutation – mistakes in DNA replication in sex cells
• CAN BE INHERITED

ENVIRONMENTAL CAUSES OF VARIATION
• Nutrients/ food • Drugs/ Chemicals • Temperature • Physical training • Light intensity • Availability of water • CANNOT BE INHERITED

A great many variations are influenced by both genetic and environmental effects. For example, your height will depend on what genes you inherit, the amount of food eaten during childhood growing period. You may inherit a good physique but unless you exercise you will not develop to your full potential. You may inherit the genes to make the pigment, melanin, in your skin, but your exposure to sunlight will darken your colour.

REPRESENTING VARIATION - GRAPHS:

Feature	Continuous Variation	Discrete/ Discontinuous Variation
Description	Variation that can be measured: - It is **quantitative**. - There are no distinct categories	Variation that can be grouped: - It is descriptive/ **qualitative**. - There are distinct categories
Examples	• Height • Weight • Hand span • Length of any body part • Hair length • Skin colour • Eye colour • Leaf size • Plant heights	• Male/female • Attached/ Attached ear lobes • Tongue roller/ non-tongue roller • Blood groups • Cattle – horns/ no horns • Colour blindness ⎫ • Albinism ⎬ Genetic • Dwarfism ⎪ defects • Sickle cell anaemia ⎭
Number of genes controlling variation	**Many genes** control these characteristics. (Polygenic inheritance)	**A single gene** controls the presence of these characteristics.
Role of the environment	Environment directly impacts the phenotype	Environment has no impact on the phenotype/ Phenotype cannot be altered
Graph used to represent type of variation	 A **HISTOGRAM** – Bars must touch as data is continuous. If sample size is large, and the midpoints of each bar are joined by a curve – a smooth bell shaped curve (a normal distribution) can be obtained. Most people are average height and few are very short or very tall in height.	 A **BAR CHART** - Bars do not touch, there is no in-between. The figures cannot be made to fit a smooth curve because there are no intermediates.

NAME:_____CLASS:_____   SKILL: ORR

TOPIC: SPECIATION   SYLLABUS OBJECTIVE: C 3.2, 4.1, 4.2, 5.1

31.   FIELD TRIP TO THE EMPEROR VALLEY ZOO – INVESTIGATING BIODIVERSITY, VARIATION
         AND SPECIATION

Do you have relatives at the zoo? Maybe the apes, but not the monkeys! A visit to the zoo is a good way to view and experience the diversity of local animals and their habitats. Some zoo exhibits include animals from other countries. It is interesting to observe what adaptations each animal demonstrates and to find out how similar looking animals belong to different species.

On this field trip, you will need to record a few pages of observations and take many pictures in order to create a report (Word Document, PowerPoint Presentation, Video, etc.) by the end of the week. Generally, scientists keep track of their observations with writings and sketches in **field journals/notebooks**. This helps them reference and remember details at a later date or allow for meaningful research to be done.

Keep in mind that your primary goals according to the syllabus are to:

✓ Identify and showcase local examples of biodiversity in the Caribbean/ your country
✓ Discuss why genetic variation is important
✓ Understand how a species is defined using either morphology or the Biological Species Concept.
✓ Observe how a particular species may be adapted to their specific habitat.

TASKS: *Obtain and record the following information in your field notebook:*

1. General information:

Date:		Duration of visit:	
Location:		Weather:	

2. Obtain a **map of the zoo**/ do a quick sketch of the layout of exhibits.
3. Locate on the map the following groupings of animals (each has two or more different species):

1. Caiman 2. Toucan 3. Cats – big vs. medium 4. Apes 5. Capuchin monkeys

4. For each animal listed above, move to the relevant exhibit(s) and observe and record:
    a. The number of different species. (Name and compare them in a table.)
    b. Physical and behavioural features that allow these different species to be grouped together.
    c. Physical, behavioural, geographical and **biological features** distinguishing the different species.
        SAMPLE TABLE TO RECORD OBSERVATIONS:

COMMON NAME	SCIENTIFIC NAME (*Genus species*)	PHYSICAL FEATURES and ADAPTATIONS	BEHAVIOURAL FEATURES (AND ADAPTATIONS)	WORLDWIDE LOCATION AND HABITAT	FOOD SOURCE(S)	SKETCH (PICTURE)

5. Include a reflection of your <u>most memorable experience</u> and <u>new learnings</u>.

ZOO FIELD TRIP – NOTES

Biodiversity is the variety of organisms found in different ecosystems. The **Zoo** exhibits a variety of animals, habitats and *ex-situ* conservation measures (outside of natural habitats and ranges). Some zoos are involved in wildlife rehabilitation, breeding and re-introduction programmes.

Classification groups living things based on features they have in common. The zoo showcases the animal **kingdom** and vertebrates belonging to the **phylum** chordata (having a brain and spinal cord surrounded by a skull and backbone). These organisms are further subdivided into **classes** based on body covering (fur on mammals, feathers on birds, dry scales on reptiles, wet scales on fish, and, skin on amphibians), reproduction and thermoregulation. Organisms in each class can be divided into **orders** such as carnivores, primates, etc. Orders are divided into **families** and then **genera** (singular **genus**) and finally **species** is the most specific level of classification.

A species may be defined as a group of organisms sharing similar physical characteristics (the **morphological species definition**). More accurately the **biological species concept** focuses more on evolutionary history and reproductive ability, where a species is *"a group of organisms with <u>a common ancestor</u> that are <u>able to interbreed</u> or mate in natural conditions and <u>have fertile offspring"</u>*, i.e. offspring that could themselves produce offspring. For example, all dogs, no matter the breed (size, height and shape) are able to successfully interbreed. A biologically defined species cannot produce fertile offspring with any other species, for example a sheep and a goat cannot produce offspring. Hybrids such as the mule (offspring of a donkey and horse) are sterile and is not a new species either.

Speciation is the formation of a new species in space or time. Usually a single species divides into two or more genetically distinct ones. It can occur by **geographic isolation** of a population from the original population due to a barrier like a mountain or river, or by colonising a new island. The new population adapts to the new environment and subsequently are unable to interbreed with the original population. New species can also be formed due to **reproductive isolation**. For example, members of the same genus of lizard, *Anolis,* occupied different locations (ecological niches) and eventually became separate species. Different behaviour (mating rituals and songs) in birds are possible barriers to reproduction.

Natural Selection is the process whereby organisms that are better adapted to a particular environment, tend to survive and have more offspring. Natural selection may lead to speciation.

Adaptations are inherited **behavioural or physical characteristics that aid survival** in the natural environment. For example, grasping tail of monkeys, opposable thumbs in humans and the sharp claws and teeth in carnivores. Not all adaptations are inherited, some randomly occur by chance due to DNA mutations. Adaptations that give a **competitive edge** allows the organism to survive and reproduce, passing on this advantageous feature to the offspring. For more info: http://evolution.berkeley.edu/.

Comparison of apes and monkeys

FEATURE	APES	MONKEYS
Intelligence	Very intelligent, can learn sign language.	Not as intelligent, cannot learn sign language
Tails	Has no tails	Most have tails that can grasp and hold
Noses	Smooth and flat	Project forward into a snout
Posture and movement	Upright and can walk on two legs. Can use arms to swing from branches.	Use all limbs for climbing, running along branches and leaping.
Examples	Gorillas, chimpanzees and orangutans	Red howler and capuchins monkeys, mandrill.

NAME:_____CLASS:_____ SKILL: AI/MM

TOPIC: NUTRITION/ ENZYMES   SYLLABUS OBJECTIVE: B 2.8, 2.9

32.   INVESTIGATING HOW pH OF A SOLUTION AFFECTS ENZYME ACTIVITY

AIM:_____

_____

Right now, in your body right there are thousands of enzymatic reactions happening. Without enzymes, life would be impossible; metabolic reactions will be too slow. A healthy digestive system has digestive enzymes to break down food into products for energy, to form new cells and grow. Enzymes are proteins that work best at optimums of temperature, pH and substrate concentration. In this experiment the ability of amylase enzyme to breakdown starch into maltose/glucose will be investigated using different pH environments.

APPARATUS AND MATERIALS:

- 5 test tubes
- Test tube rack
- Spotting tile
- 2 Measuring cylinders
- 4 Droppers
- Syringe
- Stop clock

- Distilled water
- Universal indicator paper
- Labels
- 1% Starch solution
- 1% Amylase solution
- Dilute acid (0.5M) – vinegar/ethanoic acid

- Dilute alkali (0.5M) – sodium hydrogencarbonate ($NaHCO_3$)
- Strong acid – hydrochloric acid (HCl)
- Iodine solution

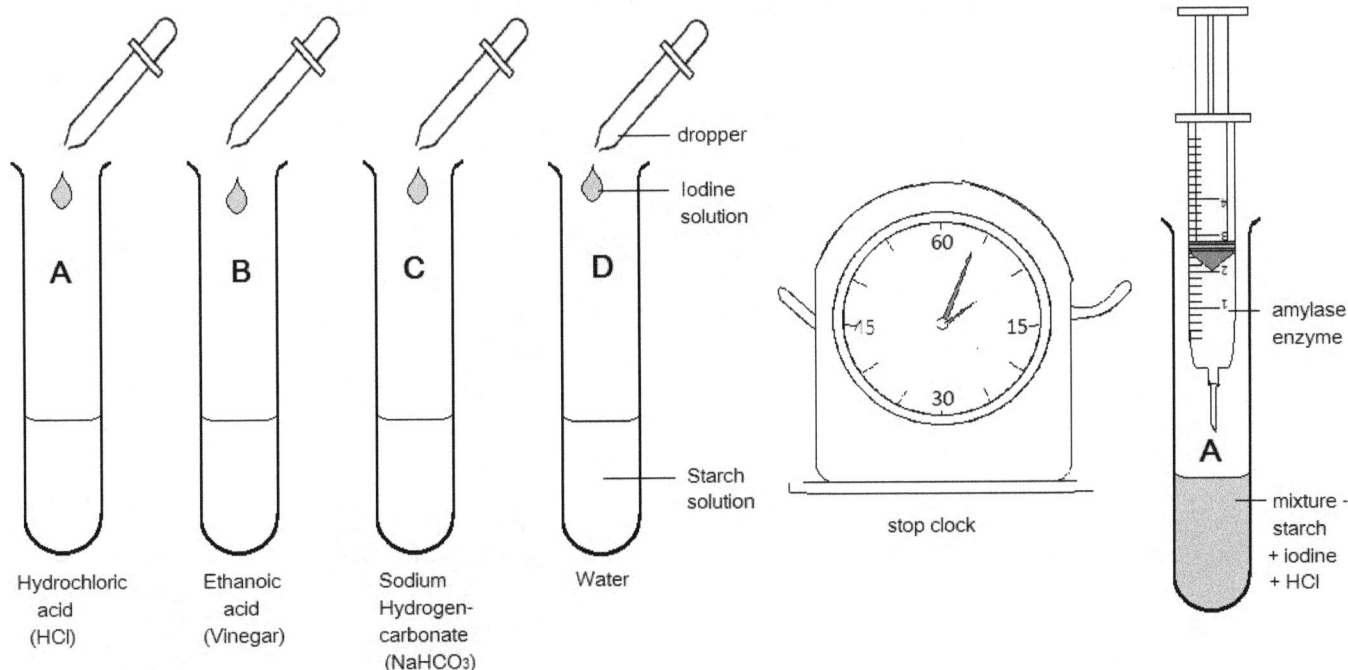

DRAWING SHOWING APPARATUS USED TO SHOW SET-UP THE EXPERIMENT

INSTRUCTIONS:

1. Label 4 test tubes A, B, C and D.
2. Determine the pH of hydrochloric acid, vinegar, sodium hydrogen carbonate and distilled water with universal indicator paper.
3. To each test tube add $5cm^3$ of starch solution.
4. To test tube A, add 2 drops of hydrochloric acid.
5. To test tube B, add 2 drops of vinegar (ethanoic acid).
6. To test tube C, add 2 drops of sodium hydrogen carbonate.
7. To test tube D, add 2 drops of water.
8. To each test tube, add one drop of iodine; shake the tubes gently to mix.
9. Note the colours of all of these test tubes in an appropriate table.
10. Measure and add $2cm^3$ amylase to test tube A; immediately start the stop clock. Record how long it took for the blue-black colour to disappear.
11. Repeat step 10 for the other test tubes, up to a maximum of 10 minutes.
12. Convert time to rate using the formula (rate = 1/time) and plot a graph to show the effect of pH on amylase activity.

TABLE 2 SHOWING CONTENTS OF FOUR TEST-TUBES TESTING EFFECT OF PH ON AMYLASE

Test tube	Contents added/ Reagents
A	Starch, iodine, amylase + hydrochloric acid
B	Starch, iodine, amylase + vinegar
C	Starch, iodine, amylase + $NaHCO_3$
D	Starch, iodine, amylase + water

Rewrite your method in past tense in the space below or on a separate page.

METHOD:

_____

_____

_____

_____

_____

_____

_____

_____

_____

_____

RESULTS:

TABLE SHOWING THE TIME TAKEN FOR AMYLASE TO BREAKDOWN STARCH USING DIFFERENT PH CONDITIONS

	TEST TUBES			
	A	**B**	**C**	**D**
	Hydrochloric acid (HCl)	Vinegar (Ethanoic acid)	Sodium hydrogen carbonate (NaHCO$_3$)	Water (H$_2$O)
pH value				
Initial Colour with iodine (0 minutes)				
Time taken for blue black colour to disappear (minutes)				
Rate of reaction (1/time)				

In the space below, convert your time to rates.

For example:

For test tube B – vinegar:

Time taken to go colourless = 4 minutes 45 seconds
         = 4.75 minutes

Then, rate of reaction $= \frac{1}{4.75}$

      = 0.21 (2 d.p.)

Convert Seconds to Minutes:

60 seconds = 1 minute

1 second = 1/60 minutes

45 seconds = 45 x (1/60) minutes

         = 45/60 minutes

      = 0.75 minutes

Space for sample calculations:

_____

_____

_____

_____

_____

_____

_____

_____

_____

Plot a graph showing the effect of pH on the breakdown of starch by amylase. Use the checklist:
- ☐ Title of graph - IN CAPITALS, below and descriptive, underlined
- ☐ Scale stated for x-axis and y-axis
- ☐ Labels on x-axis and y-axis with units stated (pH value, rate per minute)
- ☐ Plot points with single neat dot and circle Ө or X or +
- ☐ Join points only using a ruler; no extrapolation (extend lines for lower and higher pH values)

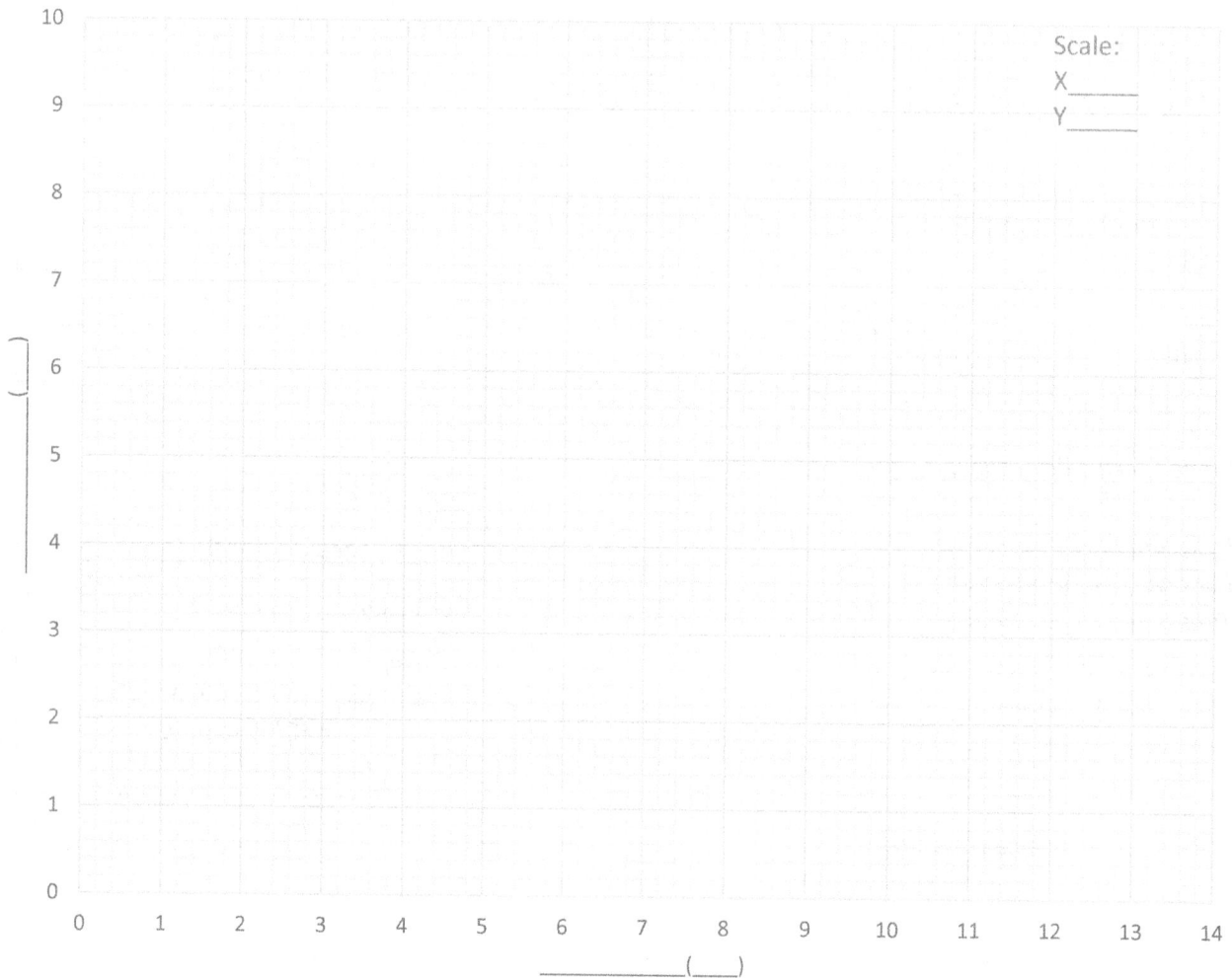

Scale:
X_____
Y_____

_____(_____)

TITLE: _____

DISCUSSION

LIMITATIONS _____

_____

_____

_____

PRECAUTIONS _____

_____

_____

_____

SOURCES OF ERROR _____

_____

_____

_____

CONCLUSION _(What was the optimum pH for amylase? Quote the values you obtained.)_____

_____

_____

_____

REFLECTION _(Do you have a better appreciation/ understanding for the conditions that affect enzyme activity. Explain.)_ _____

_____

_____

_____

_____

_____

_____

_____

_____

DISCUSSION QUESTIONS:

1. What is an enzyme? Give 3 properties of enzymes.
2. Name 2 places amylase is found in the body.
3. Write the equation for the breakdown reaction of starch due to amylase enzyme.
4. What identifies if an enzyme is working? (Hint – disappearance of substrate – starch in this case; or appearance of products – maltose/ reducing sugars.)

5. What are optimum conditions for enzymes and why is this important for enzyme function (Hint – maintaining the shape of the enzyme active site/ prevent denaturation.)
6. Identify the manipulated variable, responding variable and constant variables in this experiment.
7. Which test tube was the control experiment and why?
8. What did the results indicate about the breakdown of starch - quote values from your graph drawn, e.g. the optimum pH for amylase was _____ since the rate of reaction was fastest at _____per minute as shown on the graph.
9. Compare the observed results with what was expected?

10. What were some precautions taken?
11. Name 2 sources of error / limitations and say how they could have affected the results.
12. How could this experiment be improved? (Hint – increase the range of pH by using special buffer solutions which resist the change in pH that may occur as products form or substrates breakdown.)

NOTES – FACTORS THAT AFFECT ENZYME ACTIVITY – pH.

Enzymes are biological (organic) catalysts produced by all living cells. They speed up chemical reactions without themselves being changed. They may be found within cells or dissolved in the fluid around cells. **Enzymes are important because without them, chemical reactions would occur too slowly for life to exist**. This is because the temperature inside organisms is too low (e.g. the human body temperature is about 37°C). Only when enzymes are present to speed them up do the reactions take place quickly enough.

pH affects enzyme activity.

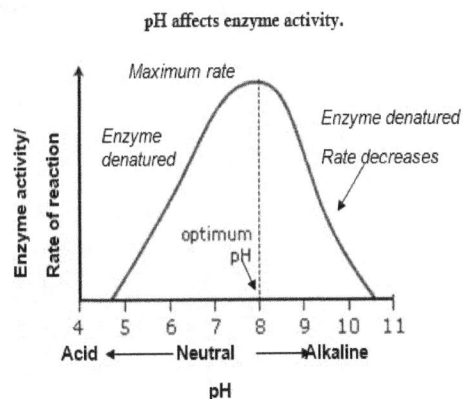

Enzyme activity and pH:pH is a measure of the acidity or alkalinity of a solution. Some enzymes work best in acidic conditions (pH<7), others work best at neutral conditions (pH =7) and other enzymes work best in alkaline conditions (pH >7). A typical graph to show how enzymes are affected by pH is shown above and described as follows:

- The optimum pH - is where the highest rate of reaction. The active site has the correct shape for the substrate to fit and break bonds.
- On either side of the optimum pH, the rate of reaction decreases as the enzyme is denatured.
- When an enzyme is denatured the active site changes shape and the substrate cannot fit into it to be broken down to form the products, thus slowing down the rate of starch break down (blue black colour disappears).

CSEC sample mark scheme for AI skill – To investigate the effect of pH on enzyme activity

AI GENERAL CRITERIA			SPECIFIC CRITERIA – To investigate the effect of pH on enzyme activity.	MAX MARK
B- Background	d s	• Define key related terms (1) • Statement of relevant theory (1)	-biological catalysts/ speed up reactions -amylase in mouth and small intestine -starch + amylase →maltose sugars	2
E - Explanation	t c u m	- Trends and patterns identified (1) - Compare actual results with expected results (1) - Use data to support explanations (1) - Modification/ improvement to existing method (1)	- The shape of the graph is --- - Expected optimum pH 7 - Observed results optimum - Comparison of Exp and Obs - Quoting values from the graph - Reference to the graph - Improvement -	4
LSP - Limitations Sources of error/ Precautions		- Identify at least 2 limitations **with explanations** - Identify at least 2 precautions/ sources of error **with explanations**	- End point could not be determined - Did not keep temperature constant - Precautions timing exactly - Source of error – contamination, inactive enzymes,	2
C- Conclusion	s r	- Statement (1) - Related to aim (1)	Optimum pH correctly identified. pH and amylase activity seen by disappearance of starch.	2
TOTAL				10

Criteria (MM) - Using the measuring cylinder + reagent bottles		Maximum Marks	Teacher Marks
DROPPER	• Equal size drops of HCl, vinegar, NaHCO and water added to the appropriate test tube	1	
	• Using a clean dropper each time a new solution added	1	
	• Equal size iodine drop placed in each test tube	1	
SYRINGE	• Plunger fully depressed at the start	1	
	• Accurately measuring 2cm^3 of amylase	1	
STOP CLOCK	• Start timing as soon as amylase added to test tube	1	
	• Correctly reading and interpreting time on stop clock	1	
	• Zero/ restart stop clock before doing next test tube	1	
PREPARATION	• Read the experiment and followed steps correctly	1	
CLEANLINESS	• Clean up work station and put away stools	1	
TOTAL		**10**	

NAME:_____   FORM:_____   SKILL: AI____

TOPIC: NUTRITION/ ENZYMES   SYLLABUS OBJECTIVE: B 2.8, 2.9

33.   INVESTIGATING HOW SUBSTRATE CONCENTRATION AFFECTS ENZYME ACTIVITY

AIM:_____

_____

Without enzymes, metabolic reactions will be too slow. Both animals and plants have enzymes in many of their tissues. Catalase is an enzyme found in potato tissue, and can be used to breakdown toxic/ corrosive hydrogen peroxide (the substrate) into harmless water and oxygen. In this fun experiment, you will measure the effect of increasing substrate concentration on the enzyme activity of catalase by recording the products formed (bubbles).

APPARATUS AND MATERIALS:

- Hydrogen peroxide (H_2O_2) of concentrations (0.5, 1.0, 1.5 and 2.0 mol/dm³ OR 1, 2, 3%)
- Potato tuber
- Distilled water
- Cork borer
- Forceps
- Ruler
- Boiling tube
- Test tube

- Delivery tube
- Rubber bungs
- Beaker of water
- Beaker for waste
- Syringe
- Stopclock
- Labels
- Blender* (optional)

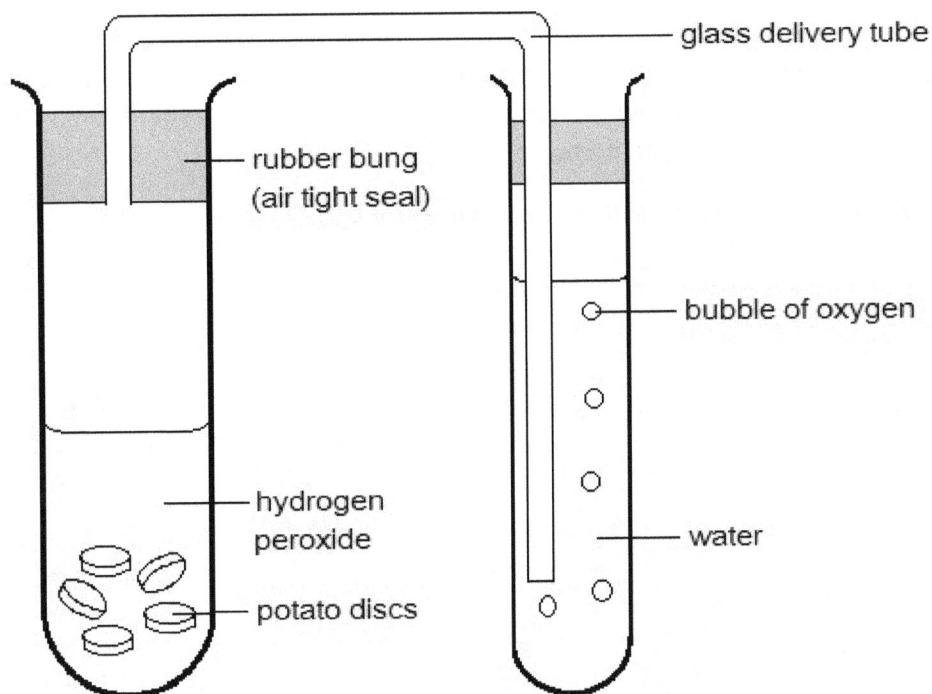

DRAWING SHOWING APPARATUS USED TO SHOW SET-UP OF APPARATUS

_____

INSTRUCTIONS:

USING POTATO DISCS

1. Use a cork borer to cut about 6 potato cylinders about 1cm in diameter and 6 cm long. Cut off peel.
2. Slice discs of 2mm thick using the knife/ scalpel and the ruler. Immediately place the discs in the beaker of water. You will require at least 60 discs.
3. Using a syringe, place 10cm³ of 0.5mol.dm³ hydrogen peroxide solution into a boiling tube. Replace the bung on the boiling tube. Set up a second tube with tap water at a depth where the delivery tube would be about 2cm below the surface of the water.
4. Count out 5 discs, and put all into the boiling tube at the same time. Replace the bung immediately, and gently shake the boiling tube to separate the slices.
5. Wait 10 seconds, and then count the number of bubbles in 1 minute. Record in a table.
6. Discard the contents of the boiling tube, and rinse it with tap water. Repeat step 4 using 5 fresh discs and hydrogen peroxide.

7. Repeat steps 3 to 6 using the other hydrogen peroxide concentrations, taking two readings each time and recording the results in the appropriate table columns.
8. Calculate the mean number of bubbles evolved for each concentration of hydrogen peroxide.
9. Plot a graph of the average number of bubbles observed (enzyme rate) using the different concentrations of substrate hydrogen peroxide.

USING BLENDED POTATO

1. Peel a medium sized potato, cut into cubes and blend with an equal volume of water at high speed.
2. Allow the mixture to settle and collect the liquid above – the homogenate (this contains the catalase enzyme).
3. Use 5 cm³ of homogenate to 5cm³ hydrogen peroxide, instead of steps 3 and 4 above.
4. After 10 seconds of mixing the solutions, count and record the number of bubbles per minute.
5. Repeat steps 3 and 4 using fresh solutions each time to get an average rate of reaction for each substrate concentration of hydrogen peroxide.

METHOD: *Rewrite your method in past tense in the space below or on a separate page.*

RESULTS:

<u>TABLE SHOWING</u> NUMBER OF BUBBLES EVOLVED PER MINUTE FROM BREAKDOWN OF HYDROGEN PEROXIDE BY CATALASE ENZYME IN POTATO

Concentration of hydrogen peroxide (H_2O_2) (mol/ dm^3)	Bubbles evolved per minute		
	1st count	2nd count	Average
0.5			
1.0			
1.5			
2.0			

Plot a graph showing the effect of **H_2O_2 substrate concentration** *on the* **rate of enzyme catalase activity.**
Use the checklist:

☐ *Title*   ☐ *Scale*   ☐ *Labels on both axes*   ☐ *Units in label*   ☐ *Plots*   ☐ *Line*

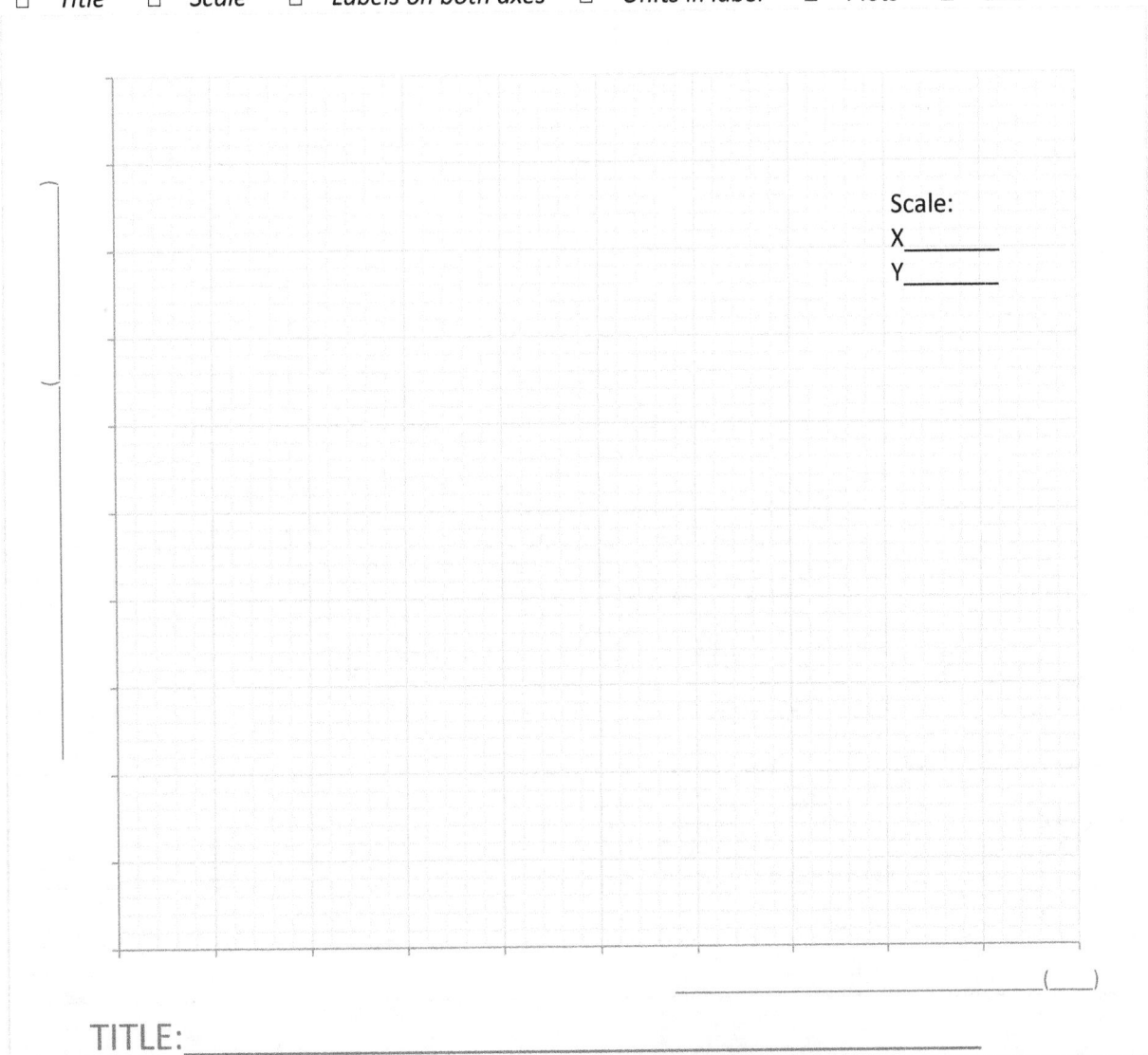

Scale:
X_____
Y_____

_____(___)

TITLE:_____

DISCUSSION

LIMITATIONS

_____

_____

_____

_____

PRECAUTIONS

_____

_____

SOURCES OF ERROR

_____

_____

_____

_____

CONCLUSION _(What was the fastest rate of reaction? Quote the values you obtained.)_

_____

_____

_____

_____

REFLECTION _(Do you have a better appreciation/ understanding for the conditions that affect enzyme activity. Explain.)_

_____

_____

_____

_____

_____

_____

_____

DISCUSSION QUESTIONS

1. What is an enzyme? Give 3 properties of enzymes.
2. Identify the enzyme and substrate in this experiment.
3. Write an equation for the breakdown of hydrogen peroxide by the enzyme in potato.
4. Why was the number of bubbles counted? (OR, height of froth, if using blended potato). Based on the equation above, what do the bubbles represent, reactant or products?

5. In general, what is the expected effect of increasing substrate concentration on enzyme activity? Explain, using a typical graph describing the active sites and saying what Vmax is.
6. Use the recorded results to describe the shape of the graph, quoting values of the highest and lowest rates of reaction (number of bubbles) as related to particular substrate concentrations.
7. Explain what was happening in the boiling tube to the rates of enzyme activity in terms of active sites of the enzyme, kinetic energy, number of molecules of substrate and enzymes.

8. Did the observed (recorded) results follow the expected results? Explain why they did or did not by explaining how any sources of error changed the expected results.

9. Explain why the following precautions were taken:
   a. Using 5 new discs (or fresh potato solution) each time
   b. Using discs (or potato solution) instead of a cylinder of potato.
   c. Renewing the hydrogen peroxide each time.
10. Identify any other precautions (air tight seal with the rubber bung and delivery tube, cutting discs to the same 2mm thickness, using discs from the same potato, etc...)
11. How could this experiment be improved? (Hint – increasing the number of counts; keeping hydrogen peroxide in the dark, maintaining a constant temperature.)

NOTES – FACTORS THAT AFFECT ENZYME ACTIVITY – SUBSTRATE CONCENTRATION

As substrate concentration increases, the rate of reaction increases. This is simply because there are more substrate molecules available to collide with active sites of the enzyme molecules present.

- At low substrate concentration, collisions between enzyme and substrate molecules are few so the rate of reaction is slow.
- Increasing substrate concentration increases the reaction rate proportionally (graph is a straight line at that substrate concentration)
- Eventually if substrate concentration keeps increasing, it will become so high that all the active sites on the enzymes are occupied (saturation of active sites). At this point, the maximum rate of reaction is reached, and rate will remain constant (cannot increase anymore).

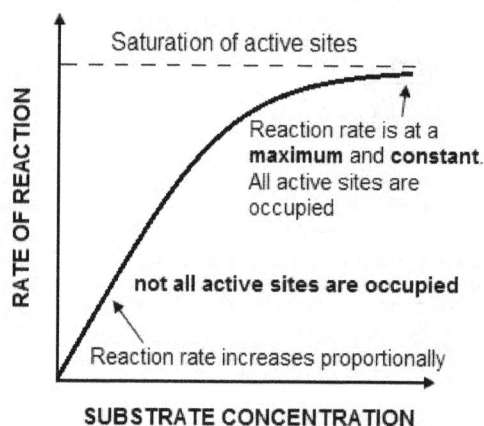

Saturation of active sites

RATE OF REACTION

Reaction rate is at a **maximum** and **constant**. All active sites are occupied

not all active sites are occupied

Reaction rate increases proportionally

SUBSTRATE CONCENTRATION

NAME:_____**CLASS:**_____ **SKILL: DR**___

TOPIC: SEXUAL REPRODUCTION IN PLANTS   SYLLABUS OBJECTIVE: B 9.7, 9.8

34.   INVESTIGATING THE FLOWER STRUCTURE OF AN INSECT POLLINATED PLANT

AIM:_____

_____

Do plants have sex? Flowers are specialized shoots that allow plants to have sex! While some flowers are male or female, most flowers are hermaphrodites, having both sets of reproductive organs   together. The stamens (male parts) produce pollen while the carpels (female parts) contain ovules. The products of plant sex is the zygote/embryo (in the seeds). Why is this important? Appreciating the basics of sexual reproduction in plants helps you understand the origins of your food like fruits and seeds. It helps highlight the importance of insects and birds in your ecosystem and even explains evolutionary adaptations as in the pea flower.

Not all flowers are the same. Some flowers are small, unscented and not colourful. They achieve pollination using the wind to transfer pollen to stigmas. However, insect pollinated flowers have colourful petals, fragrance and nectar to attract pollinators. These plants developed complex relationships with insects and even birds and bats. Among the insect pollinated flowers there is even more variation, some flowers may have parts fused together. You will investigate the general features of an insect pollinated flower in this experiment.

APPARATUS AND MATERIALS:

- Hibiscus/ Pride of Barbados flower
- Scalpel/ razor blade or knife
- White tile/ cutting surface

- Petri dish
- Forceps
- Labelling and annotation guide

INSTRUCTIONS:
1. Obtain a flower of the Hibiscus/ Pride of Barbados plant.
2. Examine its external features such as the shape, size, colour, location and arrangement of the stamen and carpels and sepals.
3. Use a knife or scalpel to cut the flower longitudinally through the ovary.
4. Make a **large, labelled drawing of the half flower**. (LABELS – petals, sepals, stigma, style, anther, filament, ovary, ovule, nectary, receptacle.)
5. Using the guide, add **at least 3** annotations (descriptions or functions of labelled parts).
6. Calculate the magnification of your drawing and state it in the title of the drawing. Use the formula:

Magnification of drawing = size of drawing/ real size of specimen (flower)

RESULTS:

Record in each box below simple, _**labelled and annotated drawings**_ of (1) the half flower and (2) separate male parts (3) separate female parts.

DRAWING CHECKLIST	
CLARITY: large	
clean/ smooth	
no shading	
ACCURACY specimen proportion	
LABELLING parallel	
accurate lower case	
justified	
annotations	
magnification	
title & view	
TOTAL	

CALCULATION OF MAGNIFICATION:

Magnification =   size of drawing
of drawing   size of specimen

= _____ ÷ _____

= X _____

TITLE:_____

CHECKLIST	
CLARITY: large	
clean/ smooth	
no shading	
ACCURACY specimen proportion	
LABELLING parallel	
accurate lower case	
justified	
annotations	
magnification	
title & view	
TOTAL	

CALCULATION OF MAGNIFICATION:

Magnification =   size of drawing
of drawing   size of specimen

= _____ ÷ _____

= X _____

TITLE:_____

Notes: Flower structure of Hibiscus– Hibiscus rosa-sinensis

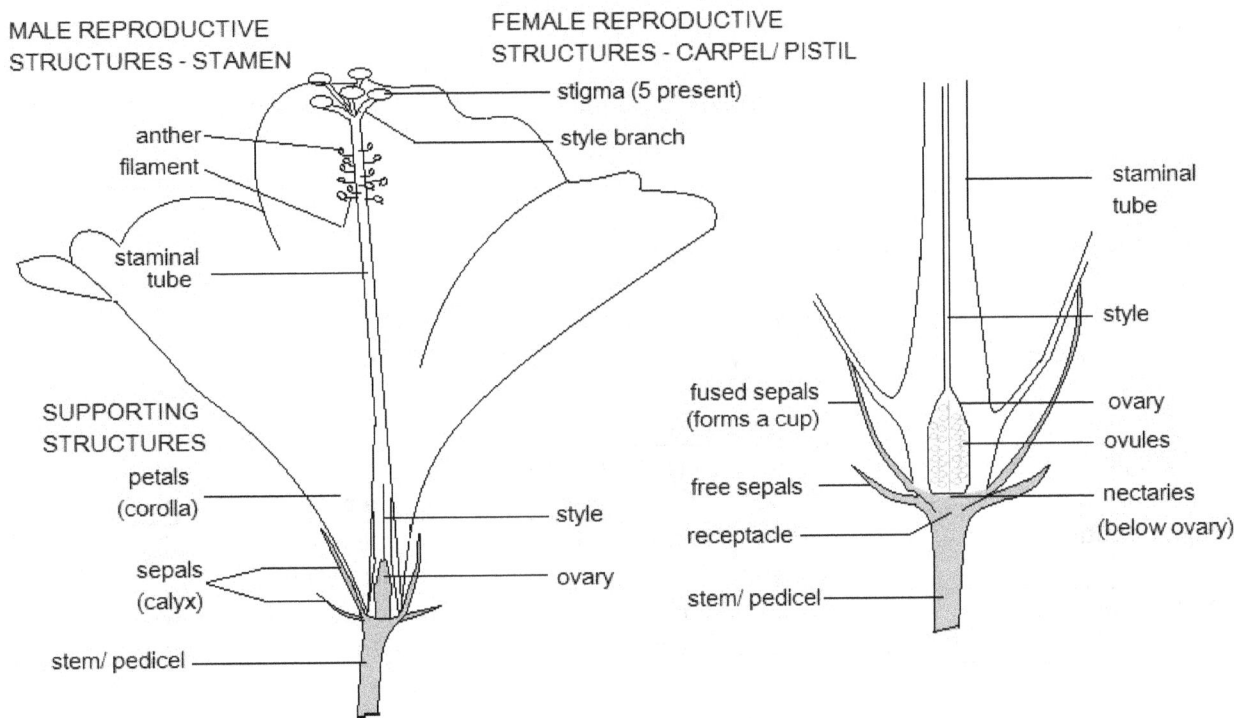

MALE REPRODUCTIVE STRUCTURES - STAMEN

FEMALE REPRODUCTIVE STRUCTURES - CARPEL/ PISTIL

- stigma (5 present)
- style branch
- anther
- filament
- staminal tube
- staminal tube
- style
- fused sepals (forms a cup)
- ovary
- ovules
- free sepals
- receptacle
- nectaries (below ovary)
- stem/ pedicel

SUPPORTING STRUCTURES
- petals (corolla)
- sepals (calyx)
- stem/ pedicel
- style
- ovary

ILLUSTRATION SHOWING THE HIBISCUS FLOWER (HALF) IN LONGITUDINAL SECTION (NOT TO SCALE) AND DETAIL OF THE OVARY AND STEM (ON THE RIGHT)

Table 1 describing the functions of the parts of the hibiscus flower.

Part		Description and/ or Function for the Hibiscus flower
Reproductive structures	Anther	Contains the pollen sacs, filled with pollen grains (inside are the male gametes).
	Filament	Supports the anther and allows it to move.
	Staminal tube	A hollow tube that - holds the anthers and filaments in place and have the stigma tube within it.
	Stigma	5 Sticky pads to attract and hold pollen grains.
	Style (inside the staminal tube).	A tube that joins the stigma to the ovary and allows pollen tubes to grow through it to deliver male gamete to the ovary.
	Ovary	Contains the ovules. The ovary becomes the fruit after the ovules are fertilized.
	Ovule	Contains the female gamete. Develops into a seed when fertilized.
Supporting structures	Petal	Large, bright colours attract pollinators and guide hummingbirds to the nectary.
	Nectary	Secretes a sweet, sugary substance to attract pollinators. Located below the ovaries.
	Sepals	Small green structures that protect the petals/flower while in bud stage. Hibiscus flowers have fused sepals form a cup and free sepals just below.
	Stem/ Pedicel	Green structure that supports the flower on the plant.
	Receptacle	Swollen top of the stem where the flower parts are attached.

COMPARISON OF INSECT POLLINATED AND WIND POLLINATED FLOWERS:

General Structures/ Parts	Insect Pollinated	Wind Pollinated
Petals	Large, brightly coloured – to attract insects	Small, dull in colour – no need to attract insects
Smell	Sweet smell – to attract insects	No scent – no need to attract insects
Nectar	Contains nectar – to attract insects	No nectar – no need to attract insects
Pollen quantity	Not much required – less wastage than with wind as insects deliver it directly to recipient stigma	Huge quantities required – most of the pollen doesn't reach another flower
Pollen characteristic	Sticky or spiky – to stick to insects	Light, dry and smooth – so it doesn't clump together and can be blown by the wind
Anthers position	Firm and inside flower – to bush against insects	Loose and outside flower – to release pollen into the wind
Stigma position	Inside flower – so that insect brushes against it	Outside flower – to catch the drifting pollen
Stigma characteristic	Sticky – so that pollen from pollinators/insects sticks to it	Sticky but also feathery and/ or net-like – to catch drifting pollen

NAME:_____**CLASS**:_____ **SKILL: DR__**

TOPIC: SEXUAL REPRODUCTION IN PLANTS   SYLLABUS OBJECTIVE: B 9.11

35.   INVESTIGATING MECHANISMS OF SEED DISPERSAL/ FRUIT DISPERSAL

AIM:_____

_____

Do plants make fruits just for you to eat? No! Fruits are actually swollen ovaries, holding the seeds – the next generation of plants. Fruits spread seeds to new locations, either by floating on water, blowing in the wind, sticking to animals, even by being excreted or just exploding and shooting seeds all around. Sometimes the seeds themselves are adapted for dispersal, leaving the fruiting body still attached to the parent plant. Other times, the whole fruits may be moved with seeds intact until deposited in a new location.

APPARATUS AND MATERIALS:

- Variety of fruits such as coconut, milkweed/ cotton, mahogany, tomato and peas pods
- White tile
- Chopping block
- Knife/cutlass
- Magnifying glass
- Labelling and identification guide – dispersal notes.

INSTRUCTIONS:
1. Obtain fruits such as coconut, milkweed/cotton/ mahogany, tomato, ochro/peas.
2. Cut the coconut, tomato and ochro/ peas longitudinally in half to expose the seeds inside.
3. Note the layers that make up the fruits and any features that help in dispersal of the fruit and the seeds within. Identify the agent of dispersal in each case.
4. Using the guide and your observations, add **at least 3** annotations (a description – size, colour, shape, texture **or** functions of labelled parts).
5. Calculate the magnification of each drawing and state it in the title of the drawing. Use the formula:

Magnification of drawing = size of drawing/ real size of specimen

RESULTS:

Record in each box below the fruits with different dispersal mechanisms. Annotate specific features that help with the type of dispersal. ***Use additional paper as needed.***

DRAWING CHECKLIST	
CLARITY: large	
clean/ smooth	
no shading	
ACCURACY specimen proportion	
LABELLING parallel	
accurate lower case	
justified	
annotation	
magnificati	
title & view	
TOTAL	

CALCULATION OF MAGNIFICATION:

Magnification = $\frac{\text{size of drawing}}{\text{size of specimen}}$
of drawing

= _____ ÷ _____

= ___X_____

TITLE:_____

DRAWING CHECKLIST	
CLARITY: large	
clean/ smooth	
no shading	
ACCURACY specimen proportion	
LABELLING parallel	
accurate lower case	
justified	
annotation	
magnificati	
title & view	
TOTAL	

CALCULATION OF MAGNIFICATION:

Magnification = $\frac{\text{size of drawing}}{\text{size of specimen}}$
of drawing

= _____ ÷ _____

= ___X_____

TITLE:_____

DISPERSAL NOTES:

Fruits protect and disperse seeds.

1. How do fruits help seed dispersal?

> The function of the fruit is to protect the seeds inside it until they are ripe/mature, and then to help disperse/spread the seeds away from the parent plant.

2. Why is seed dispersal important?

> Dispersal of seeds is important, because
> 1. It prevents too many plants growing close together. If this happens, they compete for light water, nutrients and space, so none of them can grow properly.
> 2. Dispersal also allows the plant to colonise (take over) new areas.

Fruits are ovaries after fertilization

1. What is the biological definition of a fruit?

> A fruit is the ovary of a flower after fertilisation, which contains seeds.

2. Are all fruits sweet?

No, fruits such as tomato, pepper and beans are commonly called vegetables.

3. How can you tell the difference between a fruit and a seed?

A fruit has **two scars**- one where it was attached to the plant (usually a fruit stem)
- a second where the style and stigma were attached.
A seed only has **one scar** – the hilum, where it was joined to the inside of the ovary.

Mechanisms of Seed Dispersal

1. **Water** – coconut (floation)
2. **Wind** – feathery hairs - milkweed/ silk cotton/ cotton/ ; winged seeds - mahogany
3. **Animal** – attractive fruits- tomato, OR use of hooks – burr grass, sweethearts
4. **Self/ Mechanical** – ochro/ peas; ruellia - explosive

WATER DISPERSAL - COCONUT

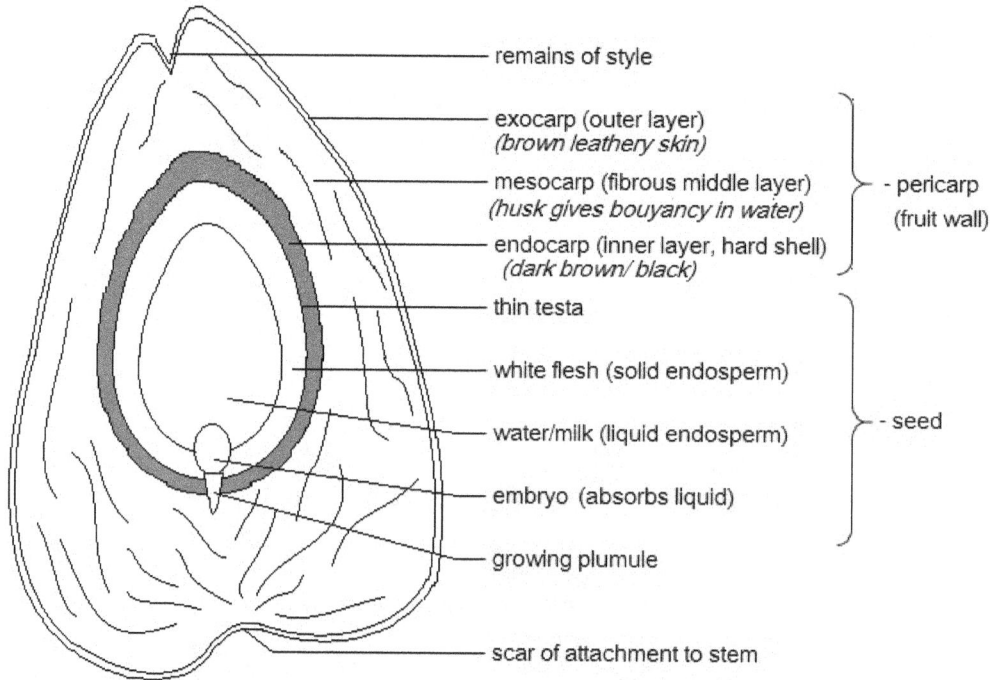

- remains of style
- exocarp (outer layer) *(brown leathery skin)*
- mesocarp (fibrous middle layer) *(husk gives bouyancy in water)*
- endocarp (inner layer, hard shell) *(dark brown/ black)*
- pericarp (fruit wall)
- thin testa
- white flesh (solid endosperm)
- water/milk (liquid endosperm)
- embryo (absorbs liquid)
- seed
- growing plumule
- scar of attachment to stem

LONGITUDINAL SECTION OF A COCONUT FRUIT (*Cocus nucifera*)

WIND DISPERSAL - A) SEEDS WITH WINGS B) SEEDS WITH HAIRS

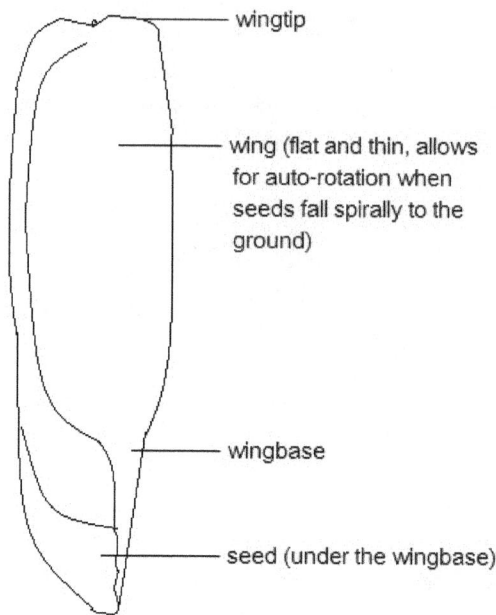

- wingtip
- wing (flat and thin, allows for auto-rotation when seeds fall spirally to the ground)
- wingbase
- seed (under the wingbase)

Mahogany seed (*Swietenia mahogani*)

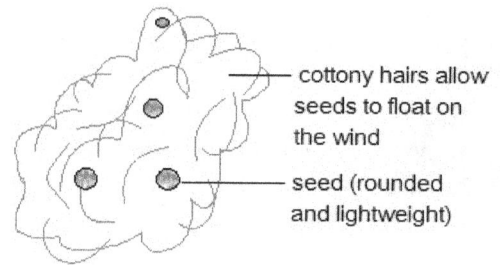

- cottony hairs allow seeds to float on the wind
- seed (rounded and lightweight)

Silkcotton seeds (*Ceiba pentandra*)

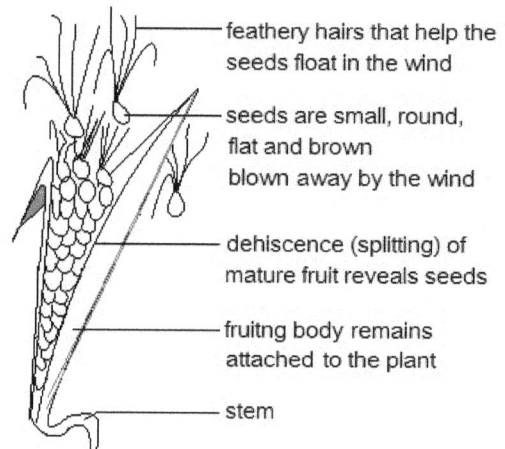

- feathery hairs that help the seeds float in the wind
- seeds are small, round, flat and brown blown away by the wind
- dehiscence (splitting) of mature fruit reveals seeds
- fruitng body remains attached to the plant
- stem

Milkweed fruit (*Asclepias curassavica*)

ANIMAL DISPERSAL

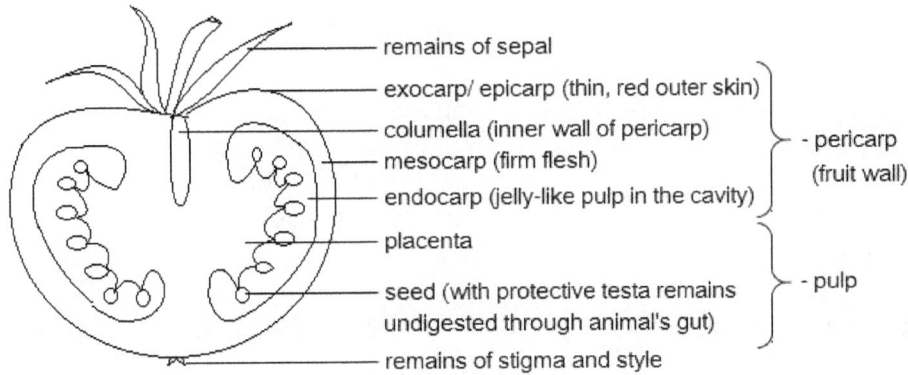

- remains of sepal
- exocarp/ epicarp (thin, red outer skin)
- columella (inner wall of pericarp)
- mesocarp (firm flesh)
- endocarp (jelly-like pulp in the cavity)
- pericarp (fruit wall)
- placenta
- seed (with protective testa remains undigested through animal's gut)
- pulp
- remains of stigma and style

Tomato fruit (berry) *(Solanum lycopersicum)*

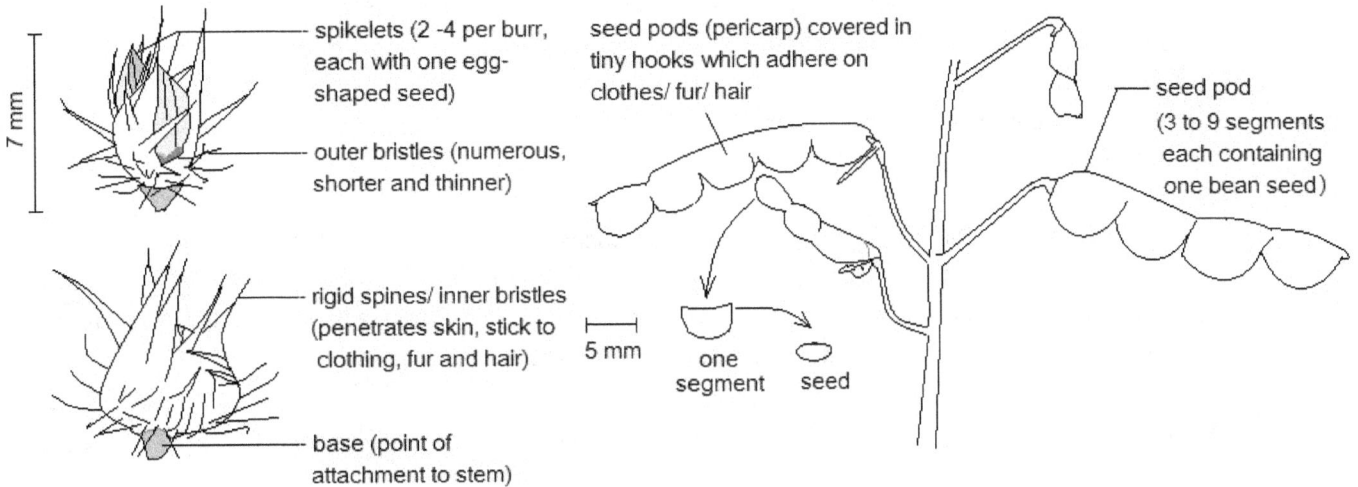

7 mm

- spikelets (2 -4 per burr, each with one egg-shaped seed)
- outer bristles (numerous, shorter and thinner)
- rigid spines/ inner bristles (penetrates skin, stick to clothing, fur and hair)
- base (point of attachment to stem)

seed pods (pericarp) covered in tiny hooks which adhere on clothes/ fur/ hair

seed pod (3 to 9 segments each containing one bean seed)

5 mm

one segment   seed

Burr grass *(Cenchrus echinatus)* spiny dry fruits   Sweethearts *(Desmodium incanum)* seedpods on a stem

SELF/ MECHANICAL DISPERSAL

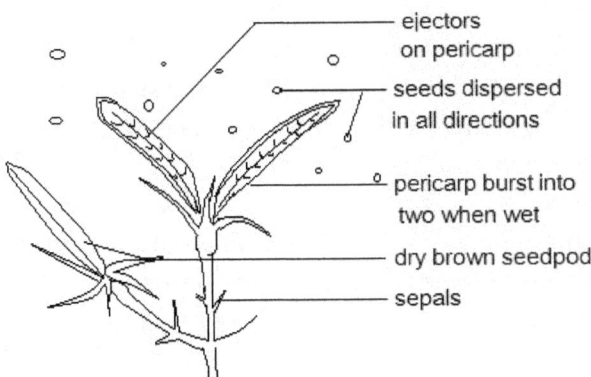

- ejectors on pericarp
- seeds dispersed in all directions
- pericarp burst into two when wet
- dry brown seedpod
- sepals

Explosive dispersal of *Ruellia tuberosa* seeds

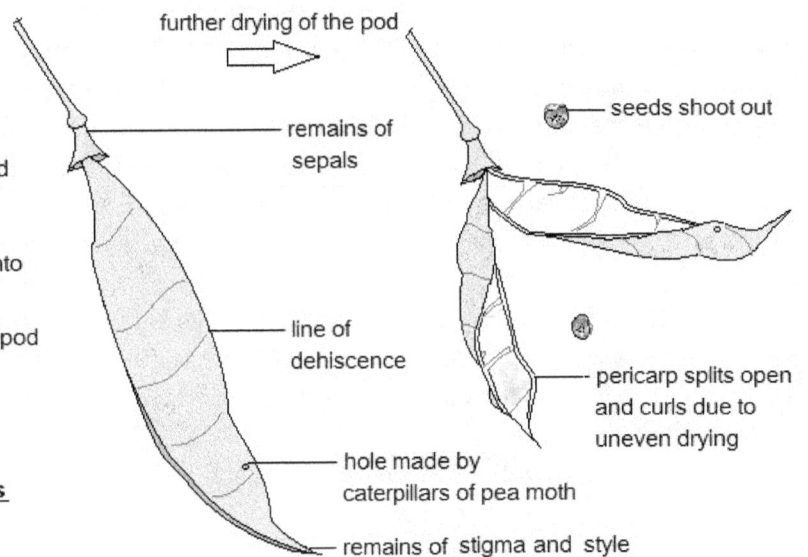

further drying of the pod

seeds shoot out

- remains of sepals
- line of dehiscence
- hole made by caterpillars of pea moth
- remains of stigma and style

pericarp splits open and curls due to uneven drying

Self/ mechanical dispersal in Pigeon peas *(Cajanus cajan)*

36.   PLAN AND DESIGN – ADDITIONAL SCENARIOS

STATEMENTS AND OBSERVATIONS USEFUL AS STIMULUS FOR PD

1. Observation: A family which keeps cows usually collects milk in the morning and leave it in their kitchen until someone has a chance to boil it and put it away in the fridge. Some days it stays out longer than others. They usually have no problems with spoiling. However, occasionally, the milk goes bad for no apparent reason. Someone suggested that it was on very hot days that it spoiled quickly but they thought this might just be a coincidence.

2. Some plants grow back faster than others after the dry season has ended.

3. Once one fruit in a bunch ripens, all the others seem to ripen very quickly after the first one.

4. Some leaves take longer than others to decompose after they have fallen.

5. Elderly people believe that putting bay leaves or neem leaves into bags of rice will keep weevils out of them.

6. Leaf-cutting ants (bachacs) seem to attack rose bushes more than other plants.

7. Students going on a field trip noticed that the boys from the football team were able to walk for a long time without getting tired while other students got tired more easily.

8. Boys cannot tell the difference between one colour and another as well as girls.

9. Yeast is a type of fungus that has many nutritional benefits when consumed. Nutritional yeast (also known as "nooch" or "yeshi") can be cultured to provide vitamin supplements in the diets of vegans and even provides a high-quality source of protein as it is forced to manufacture its own amino acids when grown in nutrient poor sugary foods like cane and beet molasses. Yeast grows best when monosaccharide sugars are present. In the past prisoners of war have used it to prevent malnutrition. Prisoners of war have access to limited food supplies such as (1) rice water (2) sugar cane juice (3) sugar cane juice boiled in dilute hydrochloric acid for 15 minutes and (4) water used to boil overripe and bruised fruit like bananas. Plan and design an experiment to determine which of those four liquids will be best for growing yeast cells.

10. In the tropical forest there are many layers of vegetation. Plants in the understory have darker green and larger leaves than those in full sun in the canopy and emergent layers. Plan and design an investigation based on the chlorophyll content of the leaves.

11. Sharon noticed that when plants in the garden were close together, some plants grew only on one side, away from the other trees.

12. Lime leaves sail more quickly on the water when playing boat races in the rain.

G. ADDITIONAL RESOURCES FOR TEACHERS

1. <u>Lab Attendance and Submission Sheet</u> – can be used when collecting labs at the end of each experiment.

LAB ATTENDANCE AND SUBMISSION SHEET

SCHOOL: _____ DATE: _____ LAB #:_____

TITLE OF LAB:_____ SKILLS:_____

CLASS	First Name	Last Name	Attendance/comments	Student signature	Marks	Marks
1						
2						
3						
4						
5						
6						
7						
8						
9						
10						
11						
12						
13						
14						
15						
16						
17						
18						
19						
20						
21						
22						
23						
24						
25						

2. The <u>List of Labs Conducted and Skills Assessed – Per Term</u> is a record for each term, the lab conducted and ensures all skills are assessed each term. See also Table 1 in Section A – Guidelines from the Syllabus.

LIST OF LABS CONDUCTED AND SKILLS ASSESSED – PER TERM			
Year	**Lab #**	**TOPIC/ TITLE/ AIM**	**Skills Assessed**
Form 4 Term 1	1		
	2		
	3		
	4		
	5		
Term 2	6		
	7		
	8		
	9		
	10		
Term 3	11		
	12		
	13		
	14		
	15		
Form 5 Term 1	16		
	17		
	18		
	19		
	20		
Term 2 (Optional)	21		
	22		
	23		
	24...		

TEACHER MARK SHEET FOR SBA SAMPLE LABS BY SKILL

STUDENT NAME	SKILLS	OBSERVING, RECORDING AND REPORTING				MANIPULATION AND MEASUREMENT					DRAWING			PLAN AND DESIGN			ANALYSIS AND INTERPRETATION			
	Year	Y1	Y1	Y2	Y2	Y1	Y1	Y2	Y2	Sum (20)	Y1	Y1	Avg. (10)	Y1	Y2	Sum (20)	Y1	Y1	Y2	Y2
	Lab				Sum (20)															Sum (20)
	Topic																			
1																				
2																				
3																				
4																				
5																				
6																				
7																				
8																				
9																				
10																				
11																				
12																				
13																				
14																				
15																				

Y1 – represents form 4   Y2 – represents form 5   Sum the averages of Y1 and Y2 per skill

FINAL MARK SHEET FOR ALLOCATION OF SBA MARKS FOR CSEC

School :   Teacher: Year of exam:

STUDENT'S NAME (Surname first Alphabetical order)	ORR (P3) TOTAL (20 marks)	DR (P3) TOTAL (10 marks)	MM (P3) TOTAL (20 marks)	PD (P3) TOTAL (20 marks)	AI (P2) TOTAL (20 marks)	TOTALS P3 (70 marks)	TOTALS P2 (20 marks)	GRAND TOTAL (90 marks)	COMMENTS (Estimated grade and Rank)
1									
2									
3									
4									
5									
6									
7									
8									
9									
10									
11									
12									
13									
14									
15									

www.ingramcontent.com/pod-product-compliance
Lightning Source LLC
Chambersburg PA
CBHW051210200326
41519CB00025B/7063